COGNITIVE ECONOMY

COGNITIVE
ECONOMY

THE ECONOMIC DIMENSION OF
THE THEORY OF KNOWLEDGE

NICHOLAS RESCHER

UNIVERSITY OF PITTSBURGH PRESS

Published by the University of Pittsburgh Press, Pittsburgh, Pa., 15260
Copyright © 1989, University of Pittsburgh Press
All rights reserved
Feffer and Simons, Inc., London

Library of Congress Cataloging-in-Publication Data

Rescher, Nicholas.
  Cognitive economy: an inquiry into the economic dimension of the theory of knowledge / Nicholas Rescher.
    p. cm.
  Includes index.
  ISBN 0-8229-3617-8
  1. Knowledge, Theory of—Economic aspects. 2. Research—Cost effectiveness. 3. Cost effectiveness—Philosophy. I. Title.
BD161.R469 1989
001—dc19   89-5425
                                                             CIP

CONTENTS

Preface   ix

ONE   KNOWLEDGE AND SCEPTICISM IN ECONOMIC PERSPECTIVE

  The Economic Dimension   4
  The Benefits of Knowledge   6
  The Costs of Knowledge   9
  Cost Effectiveness as a Salient
    Aspect of Rationality   11
  Attitudes to Risk   14
  Scepticism and Risk Aversion   17
  Why Cognitive Risks Are Worth Running   22
  Against Scepticism   29

TWO   THE ECONOMICS OF TRUST AND COOPERATION

  The Cost Effectiveness of Sharing and
    Cooperating in Information Acquisition and
    Management   33
  The Advantages of Cooperation   36
  Building up Trust: An Economic Approach   38
  A Community of Inquirers   43

THREE   ECONOMIC ASPECTS OF COMMUNICATION

  Communication Requires Conceding and
    Maintaining Credibility   47

Communication as a General Benefit
   Enterprise   51
Aspects of Scientific Communication   58
Scientific versus Ordinary-Life
   Communication   62

FOUR   IMPORTANCE AND ECONOMIC RATIONALITY

The Assessment of Importance   69
Cognitive Importance as a Typical Case   72
The Role of Importance in Economic
   Rationality   80

FIVE   INDUCTION, SIMPLICITY, AND COGNITIVE ECONOMY

Induction as Cognitive Systematization   83
Induction and Cognitive Economy:
   The Economic Rationale of Simplicity
   Preference   88
The Methodological Aspect of Inductive
   Economy   95
The Ontological Ramifications of
   Simplicity   100

SIX   ECONOMICS AND THE METHODOLOGY OF INQUIRY

Introduction   109
Hempel's Paradox of the Ravens   109
Goodman's Grue Paradox   113
Generality Preference and Falsificationism   118
Novelty Tropism   123
Making versus Postponing Decisions   124

CONTENTS

Symmetry Arguments   126
Conclusion   129

SEVEN COST ESCALATION IN EMPIRICAL INQUIRY

The Exploration Model of Scientific Inquiry   132
Technological Escalation   134
Theorizing as Inductive Projection   138
The Dialectic of Experimentation and
   Theorizing   141
The Economic Limits of Science   143
Conclusion   150

Notes   155

Name Index   167

PREFACE

The American polymath Charles Sanders Peirce (1839–1914) was the first theorist to stress the role of economic considerations in the theory of knowledge. The project of an "economy of research" to which he adverted in various sketches, had it been carried to completion at his hands, would have been an elegant and compelling exposition of this idea. But, as fate would have it, all he left us were brief discussions and tantalizing observations. The present book, however imperfect, at least represents an effort to pick up the torch and pass its flame along.

The book falls into three parts. The first, consisting of the opening chapter alone, indicates how the cognitive project at large can be viewed in economic perspective. The second, consisting of chapters 2 and 3, examines the communal and cooperative aspect of inquiry as firmly rooted in cost-effectiveness considerations. The third and longest part, comprising the remaining four chapters, examines various economic aspects of the scientific enterprise in particular. Overall, the book seeks to demonstrate the utility of an economic perspective for understanding the ways and means of our cognitive endeavors.

The theoretical objectives of the project constrain much of the discussion to proceed on a rather abstract plane, but the case studies presented in the two final chapters should prevent the reader from feeling that there is only theory and no application.

The book is a synthesis and consolidation of more sporadic earlier discussions of economic factors bearing on the rational conduct of inquiry. Chapter 1 is indebted to *Scep-*

ticism (Oxford, 1980); chapter 5 to *Induction* (Oxford, 1980); chapter 6 to *Peirce's Philosophy of Science* (Notre Dame and London, 1978), and chapter 7 to *Scientific Progress* (Oxford, 1978). Like most of my recent books, this one develops and systematizes ideas that played a more incidental and subsidiary role in my earlier writings.

This book was projected in the autumn of 1986, drafted in Pittsburgh during the course of that academic year, refined in Oxford during the rainy summer of 1987, and further polished in Pittsburgh during the 1987–88 academic year. In revising my manuscript, I have profited from the helpful comments of David Carey. And I am very grateful to Mrs. Linda Butera and Mrs. Christina Masucci for their help in seeing the manuscript through numerous revisions on the word processor.

Pittsburgh, Pennsylvania
May 1988

COGNITIVE ECONOMY

ONE

Knowledge and Scepticism in Economic Perspective

SYNOPSIS

(1) The economics of knowledge is an important but underdeveloped branch of epistemology. It is—or should be—evident that knowledge has its economic aspect of benefits and costs. (2) The benefits of information are both theoretical and applied. (3) Moreover, the management of information is always a matter of costs. (4) Rationality itself has a characteristically economic dimension in its insistence on a proper proportioning of expenditures and benefits. (5) People generally have very different sorts of attitudes toward cognitive risk. And the cognitive project of accepting information—of endorsing answers to our questions—is a matter of balancing risks: the risk of error and ignorance as against the benefit of knowledge, of possible opportunities forgone as against possible successes achieved. (6) Scepticism roots in an undue (it is tempting to say neurotic) aversiveness to cognitive risk. The chancing of loss for the prospect of gain—the running of risk of error to achieve the potential advantage of knowledge—is an inevitable aspect of the

human condition. (7) Overall, the sensible balance of costs and benefits in the management of cognitive risks certainly does not favor scepticism. To be sure, such an appeal to the economic aspect of cognition will doubtless not move the committed sceptic. But it should suffice to deter someone who is not already precommitted to the position from taking it in the first place.

THE ECONOMIC DIMENSION

The theme of this book is the importance of *economic* considerations in the conduct of our *cognitive* affairs. Knowledge has a significant economic dimension because of its substantial involvement with costs and benefits. Many aspects of the way we acquire, maintain, and use our knowledge can be properly understood and explained only from an economic point of view. Attention to economic considerations regarding the costs and benefits of the acquisition and management of information can help us both to account for how people proceed in cognitive matters and to provide normative guidance toward better serving the aims of the enterprise. Any theory of knowledge that ignores this economic aspect does so at the risk of its own adequacy.

It has come to be increasingly apparent in recent years that knowledge is *cognitive capital,* and that its development involves the creation of intellectual assets, in which both producers and users have a very real interest. Knowledge, in short, is a good of sorts—a commodity on which one can put a price tag and which can be bought and sold much like any other—save that the price of its acquisition often involves not just money alone but other resources, such as time, effort, and ingenuity. Man is a finite being who has only limited time and energy at his disposal. And even the development of knowledge, important though it is, is nevertheless of limited value—it is not worth the expenditure of every minute of every day at our disposal.

Charles Sanders Peirce proposed to construe the "economy of research" at issue in knowledge development in terms of the sort of balance of assets and liabilities that we today would call cost-benefit analysis.[1] On the side of benefits of scientific claims, he was prepared to consider a wide variety of factors: closeness of fit to data, explanatory value, novelty, simplicity, accuracy of detail, precision, parsimony, concordance with other accepted theories, even antecedent likelihood and intuitive appeal. And in the liability column, he placed those demanding factors of "the dismal science": the expenditure of time, effort, energy, and money needed to substantiate our claims.

The introduction of such an economic perspective does not of course detract from the value of the quest for knowledge as an intrinsically worthy venture with a perfectly valid *l'art pour l'art* aspect. But as Peirce emphasized, one must recognize the inevitably economic aspect of any human enterprise—inquiry included.

The value of knowledge is, for the purpose of science, in one sense absolute. It is not to be measured, it may be said, in money; in one sense that is true. But knowledge that leads to other knowledge is more valuable in proportion to the trouble it saves in the way of expenditure to get that other knowledge. Having a certain fund of energy, time, money, etc., all of which are merchantable articles to spend upon research, the question is how much is to be allowed to each investigation; and *for* us the value of that investigation is the amount of money it will pay us to spend upon it. *Relatively*, therefore, knowledge, even of a purely scientific kind, has a money value. (*Collected Papers*, vol. I [Cambridge, MA, 1931], sec. 1.122 [c.1896])

Philosophical epistemologists subsequent to Peirce have paid regrettably little attention to these matters.[2] Indeed, they often proceed on the tacit assumption that information is something that is economically costless—a free good that comes to rational inquirers without any expenditure and effort apart from thought itself. But even casual consideration shows that this is totally erroneous and unrealistic. Knowledge only comes to us at a price.

THE BENEFITS OF KNOWLEDGE

Knowledge brings benefits too. Man has evolved within nature into the ecological niche of an intelligent being. In consequence, the need for understanding, for "knowing one's way about," is one of the most fundamental demands of the human condition. Man is *Homo quaerens*. The requirement for information, for cognitive orientation within our environment, is as pressing a human need as that for food itself. We are rational animals and must feed our minds even as we must feed our bodies.

The need for knowledge is part and parcel of our nature. A deep-rooted demand for information and understanding presses in upon us, and we have little choice but to satisfy it. Once the ball is set rolling it keeps on under its own momentum—far beyond the limits of strictly practical necessity. The great Norwegian polar explorer Fridtjof Nansen put it well. What drives men to the polar regions, he said, is

> the power to the unknown over the human spirit. As ideas have cleared with the ages, so has this power extended its might, and driven Man willy-nilly onwards along the path of progress. It drives us in to Nature's hidden powers and secrets, down to the immeasurably little world of the microscopic, and out into the unprobed expanses of the Universe. . . . it gives us no peace until we know this planet on which we live, from the greatest depth of the ocean to the highest layers of the atmosphere. This Power runs like a strand through the whole history of polar exploration. In spite of all declarations of possible profit in one way or another, it was that which, in our hearts, has always driven us back there again, despite all setbacks and suffering.[3]

The discomfort of unknowing is a natural component of human sensibility. To be ignorant of what goes on about us is almost physically painful for us—no doubt because it is so dangerous from an evolutionary point of view. As William James observed: "The utility of this emotional effect of [security of] expectation is perfectly obvious; "natural selection," in fact, was bound to bring it about sooner or later. It is of the utmost practical importance to an animal that he should have prevision of the qualities of the objects that surround him."[4]

It is a situational imperative for us humans to acquire information about the world. We have questions and we need answers. Homo sapiens is a creature who must, by his very nature, feel cognitively at home in the world. Relief from ignorance, puzzlement, and cognitive dissonance is one of cognition's most important benefits. These benefits are both positive (pleasures of understanding) and negative (reducing intellectual discomfort through the removal of unknowing and ignorance and the diminution of cognitive dissonance). The basic human urge to make sense of things is a characteristic aspect of our make-up—we cannot live a satisfactory life in an environment we do not understand. For us, cognitive orientation is itself a practical need: cognitive disorientation is actually stressful and distressing.

The benefits of knowledge are twofold: theoretical (or purely cognitive) and practical (or applied). The theoretical/cognitive benefits of knowledge relate to its satisfactions in and for itself, for understanding is an end unto itself and, as such, is the bearer of important and substantial benefits—benefits which are purely cognitive, relating to the informativeness of knowledge as such. The practical benefits of knowledge, on the other hand, relate to its role in guiding the processes by which we satisfy our (noncognitive) needs and wants. The satisfaction of our needs for food, shelter, protection against the elements, and security against natural and human hazards all require information. And the satisfaction of mere desiderata comes into it as well.

We can, do, and must put knowledge to work to facilitate the attainment of our goals, guiding our actions and activities in this world into productive and rewarding lines. And this is where the practical payoff of knowledge comes into play.

The impetus to inquiry—to investigation, research, and acquisition of information—can thus be validated in strictly economic terms with a view to potential benefits of both theoretical and practical sorts. We humans need to achieve both an intellectual and a physical accommodation to our environs. Efforts to secure and enlarge knowledge are worthwhile insofar as they are cost effective, in that the resources we expend for these purposes are more than compensated for through benefits obtained.

THE COSTS OF KNOWLEDGE

No human activity comes free of charge; everything we do has its costs in terms of time, energy, effort, physical goods, or the like. Throughout the entire range of our endeavors in this world, we are involved in the expenditure of limited resources. And knowledge is no exception to this rule. Its acquisition, processing, storage, retrieval, and utilization are activities which, like any other human endeavor, engender costs. And over and above this practical dimension there are also certain purely cognitive disabilities and negativities—that is, costs—involved in ignorance, error, and confusion.

The problems of knowledge acquisition are certainly not without their economic ramifications. The costs (and benefits) of knowledge acquisition vary with people's conditions and circumstances. Time is of the essence here. The medical knowledge of the twentieth century was not available to patients in the eighteenth century—"not for all the tea in China." Nor does it do someone much good to know the outcome of a horse race, an election, or a battle only after the fact, when the opportunities to capitalize thereon are long past. In pursuing information, as in pursuing food, we have to settle for the best we can obtain at the time. We have questions and need answers—the best answers we can get here and now, regardless of their imperfections. We cannot wait until all returns are in. Our needs and wants impel us to resolve our questions by means of the best answers we can get here and now. What matters for us is not ideal and certain knowledge in the light of complete and perfected information but rather simply getting the best estimate of the truth that we can manage to secure here and now.

It also deserves stress in this connection that what matters is not just having a bit of knowledge but being able to get at it. My acquaintance's telephone number is unquestionably listed somewhere in the telephone book I have here, but that will do me little good if I have forgotten his name. Whenever the information I have includes vastly more than the information I need, it may cost a good deal of time and effort to find that needle in the haystack.

Even where information is available, it may be so widely scattered and disparately deployed as to be useless. One of the gravest of cognitive problems in the modern world is that of rendering accessible in an organized, coherent, and coordinated way the information already, broadly speaking, available—a process that is invariably difficult and expensive. In the processing of information, the left hand often does not know what the right hand is doing. The need to secure information about information is an important aspect of the economics of the enterprise.

COST EFFECTIVENESS AS A SALIENT ASPECT OF RATIONALITY

The ancients saw man as the rational animal (*zoōn logon echōn*), set apart from other creatures by capacities for speech and deliberation. Under the precedent of Greek philosophy, Western thinkers have generally deemed the use of thought for the guidance of our proceedings to be at once the glory and the duty of homo sapiens.

To behave rationally is to make use of one's intelligence to figure out the best thing to do in the circumstances. Rationality consists in the use of reason to resolve choices in the best feasible way. Above all, it calls for the intelligent pursuit of appropriate ends, for the effective and efficient cultivation of appropriately appreciated benefits. Rationality requires doing the best one can with the means at one's disposal, striving for the best results that one can expect to achieve within the range of one's resources, specifically including one's intellectual resources. Be it in matters of belief, action, or evaluation, its mission centers about the deliberate endeavor to maximize benefits relative to expenditures.

Accordingly, rationality has an ineliminably economic dimension. It is quite irrational to expend more resources on the realization of a given end than one needs to.[5] The optimal use of resources is a crucial aspect of rationality. It is against reason to expend more resources on the pursuit of a goal than it is worth—to do things in a more complex, inefficient, or ineffective way than is necessary in the circumstances. It is also against reason to expend fewer resources in the pursuit of a goal than it is worth, unless these resources can be used to even better effect elsewhere. Cost effectiveness—the proper coordination of costs and benefits in the pursuit of our ends—is an indispensable requisite of rationality.

And this general situation obtains with particular force where the transaction of our specifically *cognitive* business is concerned. With any source of information or method of information acquisition, two salient questions arise:

1. *Utility:* How useful is it; how often do we have occasion/need to make use of it; and how large are the issues that rest on its availability?

2. *Cost:* How costly is its employment; how expensive (complicated, difficulty, resource-demanding) is its use?

A natural tendency is at work in human affairs (and presumably in the dealings or rational agents generally) to keep these two items in alignment—to maintain a proper proportioning of costs and benefits. In particular:

1. If a resource affords a relatively inexpensive means to accomplishing a needed task, we incline to make more use of it.

2. If we need to achieve a certain end often, then we try to devise less expensive ways of achieving this end.

Such principles of economic rationality not only explain why people use more staples than paper clips, but also account for important cognitive situations—for example, why the most frequently used words in a language tend to be among the shortest. (No ifs, ands, or buts about that!)

It is particularly noteworthy from such an economic point of view that there will be some conditions and circumstances in which the cost of acquiring information—even assuming that it is to be had at all in the prevailing state of the cognitive art—is simply too high relative to its value. There are (and are bound to be) circumstances in which the acquisition costs of information exceed the benefits or returns on its possession. In this regard, too, information is just like any other commodity. The price is sometimes more than we can afford and often greater than any conceivable benefit that would ensue. (This is why people do not count the number of hairs in their eyebrows.)

Rationality and economy are inextricably interconnected. Rational inquiry is a matter of epistemic optimization, of achieving the best overall balance of cognitive benefits relative to cognitive costs. Cost-benefit calculation is the crux of the economy of effort at issue. The principle of least effort—construed in a duly intellectualized manner—is bound to be a salient feature of cognitive rationality.[6] A version of Occam's Razor obtains throughout the sphere of cognitive rationality: *complicationes non multiplicandae sunt praeter necessitatem.*

Economy of effort is a cardinal principle of rationality that helps to explain many features of the way in which we transact our cognitive business. Why are encyclopedias organized alphabetically rather than topically? Because this simplifies the search process. Why are accounts of people's doings or a nation's transactions standardly presented historically, with biographies and histories presented in chronological order? Because an account that moves from causes to effects simplifies understanding. Why do libraries group books together by topic and language rather than, say, alphabetically by author? Because this minimizes the difficulties of search and access. We are in a better position to understand innumerable features of the way in which people conduct their cognitive business once we take the economic aspect into account.

The standard economic process of cost-effectiveness tropism is operative throughout the cognitive domain. Rational inquiry is rigorously subject to the economic impetus to securing maximal product for minimal expenditure. Concern for answering our questions in the most straightforward, most cost-effective way is a crucial aspect of cognitive rationality in its economic dimension.

ATTITUDES TO RISK

But perhaps the whole of this cognitive venture is a fraud and delusion. The scientific researcher, the inquiring philosopher, and the plain man, all desire and strive for information about the "real" world. The

Knowledge and Scepticism

sceptic rejects their ventures as vain and their hopes as foredoomed to disappointment from the very outset. As he sees it, any and all sufficiently trustworthy information about factual matters is simply unavailable as a matter of general principle. To put such a radical scepticism into a sensible perspective, it is useful to consider the issue of cognitive rationality in the light of the situation of risk taking in general.

There are three very different sorts of personal approaches to risk and three very different sorts of personalities corresponding to these approaches, as follows:

Type 1: Risk avoiders
Type 2: Risk calculators
    2.1: cautious
    2.2: daring
Type 3: Risk seekers

The type 1, risk-avoidance, approach calls for risk aversion and evasion. Its adherents have little or no tolerance for risk and gambling. Their approach to risk is altogether negative. Their mottos are Take no chances, Always expect the worst, and Play it safe.

The type 2, risk-calculating, approach to risk is more realistic. It is a guarded middle-of-the-road position, based on due care and calculation. It comes in two varieties. The type 2.1, cautiously calculating, approach sees risk taking as subject to a negative presumption, which can, however, be defeated by suitably large benefits. Its line is; Avoid risks unless it is relatively clear that a suitably large gain beckons at sufficiently suspicious odds. If reflects the path of prudence and guarded caution. The type 2.2, daringly calculating, approach sees risk taking as subject to a positive presumption, which can, however, be defeated by suitably large negativities. Its line is; Be prepared to take risks unless it is relatively clear that an unacceptably large loss threatens at sufficiently inauspicious odds. It reflects the path of optimistic hopefulness.

The type 3, risk-seeking, approach sees risk as something to be welcomed and courted. Its adherents close their eyes to danger and take a rosy view of risk situations. The mind of the risk seeker is intent on the delightful situation of a favorable issue of events: the sweet savor of success is already in his nostrils. Risk seekers are chance takers and go-for-broke gamblers. they react to risk the way an old warhorse responds to the sound of the musketry: with eager anticipation and positive relish. Their motto is; Things will work out.

In the conduct of practical affairs, risk avoiders are hypercautious; they have no stomach for uncertainty and insist on playing it absolutely safe. In any potentially unfavorable situation, the mind of the risk avoider is given to imagining the myriad things that could go wrong. Risk seekers, on the other hand, leap first and look later, apparently counting on a benign fate to ensure that all will be well; they dwell in the heady atmosphere of "anything may happen." Risk calculators take a middle-of-the road approach. Proceeding with care, they take due safeguards but still run risks when the situation looks sufficiently favorable. It is thus clear that people can have very different attitudes toward risk.

So much for risk taking in general. Let us now look more closely at the cognitive case in particular.

SCEPTICISM AND RISK AVERSION

The situation with regard to specifically cognitive risks can be approached as simply a special case of the general strategies sketched above. In particular, it is clear that risk avoidance stands coordinate with scepticism. The sceptic's line is; Run no risk of error; take no chances; accept nothing that does not come with ironclad guarantees. And the proviso here is largely academic, seeing that little if anything in this world comes with ironclad guarantees—certainly nothing by way of interesting knowledge. By contrast, the adventuresome syncretist is inclined, along with radical Popperians such as P. K. Feyerabend, to think that anything goes. His cognitive stance is tolerant and open to input from all quarters. He is gullible, as it were, and stands ready to endorse everything and to see good on all sides. The evidentialist, on the other hand, conducts his cognitive business with comparative care and calculation, regarding various sorts of claims as perfectly acceptable, provided that the evidential circumstances are duly favorable.

The sceptic accepts nothing, the evidentialist only the chosen few, the syncretist virtually anything. In effect, the positions at issue in scepticism, syncretism, and evidentialism simply replicate, in the specifically cognitive domain, the various approaches to risks at large.

It must, however, be recognized that in general two

fundamentally different kinds of misfortunes are possible in situations where risks are run and chances taken:

1. We reject something that, as it turns out, we should have accepted. We decline to take the chance, we avoid running the risk at issue, but things turn out favorably after all, so that we lose out on the gamble.
2. We accept something that, as it turns out, we should have rejected. We do take the chance and run the risk at issue, but things go wrong, so that we lose the gamble.

If we are risk seekers, we will incur few misfortunes of the first kind, but, things being what they are, many of the second kind will befall us. On the other hand, if we are risk avoiders, we shall suffer few misfortunes of the second kind, but shall inevitably incur many of the first. The overall situation has the general structure depicted in figure 1.

Clearly, the reasonable thing to do is to adopt a policy that minimizes misfortunes overall. It is thus evident that both type 1 and type 3 approaches will, in general, fail to be rationally optimal. Both approaches engender too many misfortunes for comfort. The sensible and prudent thing is to adopt the middle-of-the-road policy of risk calculation, striving as best we can to balance the positive risks of outright loss against the negative ones of lost opportunity. Rationality thus counterindicates approaches of type 1 and type 3, taking the line of the counsel, Neither avoid nor court risks, but manage them prudently in the search for an overall minimization of misfortunes. The rule of reason

FIGURE 1
Risk Acceptance and Misfortunes

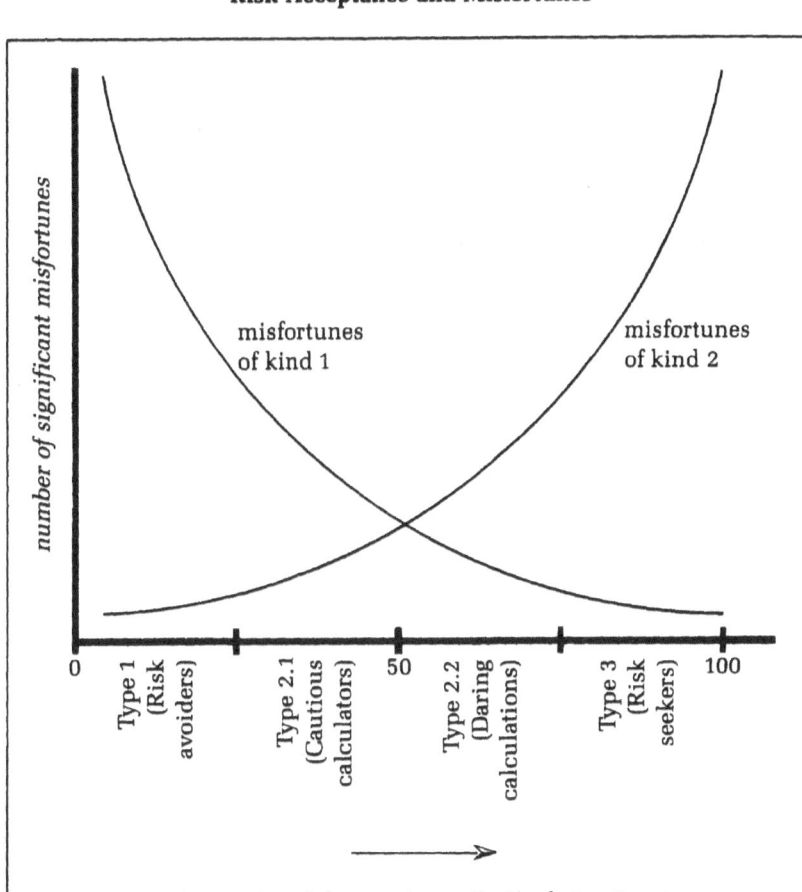

calls for sensible management and a prudent calculation of risks; it standardly enjoins upon us the Aristotelian golden mean between the extremes of risk avoidance and risk seeking.

Turning now to the specifically cognitive case, it may be observed that the sceptic succeeds splendidly in averting misfortunes of the second kind. He make no errors of commission; by accepting nothing, he accepts nothing false. But, of course, he loses out on the opportunity to obtain any sort of information. The sceptic thus errs on the side of safety, even as the syncretist errs on that of gullibility. The sensible course is clearly that of a prudent calculation of risks.

Being mistaken is unquestionably a negativity. When we accept something false, we have failed in our endeavors to get a clear view of things—to answer our questions correctly. And moreover, mistakes tends to ramify, to infect environing issues. If I (correctly) realize that P logically entails Q but incorrectly believe not-Q, then I am constrained to accept not-P, which may well be quite wrong. Error is fertile of further error. So quite apart from practical matters (suffering painful practical consequences when things go wrong), there are also the purely cognitive penalities of mistakes—entrapment in an incorrect view of things. All this must be granted and taken into account. But the fact remains that errors of commission are not the only sort of misfortune there are.[7] Ignorance, lack of information, cognitive disconnection from the world's course of things—in short, errors of omission—are also negativities of substantial proportions. This too is something we must work into our reckoning.

In claiming that his position wins out because it makes the fewest mistakes, the sceptic uses a fallacious system of scoring, for while he indeed makes the fewest errors of one kind, he does this at the cost of proliferating those of another. Once we look on this matter of error realistically, the sceptic's vaunted advantage vanishes. The sceptic is simply a risk avoider, who is prepared to take no risks and who stubbornly insists on minimizing errors of the second kind alone, heedless of the errors of the first kind into which he falls at every opportunity.

The sceptic inclines to overlook the fact that what we want in inquiry—the *raison d'être* of the whole enterprise—is information. What we seek is the very best overall balance between answers to our questions and ignorance or misinformation. We face a trade-off at this stage, however. Are we prepared to run a greater risk of mistakes to secure the potential benefits of greater understanding? The judicious cognitivist is a risk calculator who recognizes the value of understanding and is prepared to gamble for its potential benefits. H. H. Price has put the salient point well:

"Safety first" is not a good motto, however tempting it may be to some philosophers. The end we seek to achieve is to acquire as many correct beliefs as possible on as many subjects as possible. No one of us is likely to achieve this end if he resolves to reject the evidence of testimony, and contents himself with what he can know, or have reason to believe, on the evidence of his own firsthand experience alone. It cannot be denied that if someone follows the policy of accepting the testimony of others unless or until he has specific reason for doubting it, the results will not be all that he might wish. Some of the beliefs which he will thereby acquire will be totally incorrect, and others partly incorrect. In this sense, the policy is certainly a risky one. . . . But it is reasonable to take this risk, and unreasonable not to take it. If we refuse to take it, we have no prospect of getting answers, not even the most tentative ones, for many of the questions which interest us.[8]

What Price says here about the testimony of others is equally true when applied to the testimony of our own senses—and indeed holds good for our resources of inquiry in general.

Ultimately, we face a question of value trade-offs. Are we prepared to run a greater risk of mistakes to secure the potential benefit of an enlarged understanding? In the end, the matter is one of priorities—of safety as against information, of ontological economy as against cognitive advantage, of an epistemological risk aversion as against the impetus to understanding. The ultimate issue is one of values and priorities, weighing the negativity of ignorance and incomprehension against the risk of mistakes and misinformation.

WHY COGNITIVE RISKS ARE WORTH RUNNING

Scepticism not only runs us into practical difficulties by impeding the cognitive direction of action, but it has great theoretical disadvantages as well. The trouble with scepticism is that it treats the avoidance of mistakes as a paramount good, one worth purchasing even at a considerable cost in ignorance and lack of understanding. For the radical sceptic's seemingly high-minded insistence on definitive truth, in contradistinction to merely having reasonable warrant for acceptance—duly followed by the mock-tragic recognition that this is of course unachievable—is totally counterproductive. It blocks from the very outset any prospect of staking reasonable claims to information about the ways of the world. To be sure, the averting of errors of commission is a very good thing when it comes free of charge, or at any rate cheap. But it may well be bought at too high a cost when it requires us to accept massive sacrifices in for going the intellectual satisfaction of explanation and understanding. It would of course be nice if we could separate errors of commission from errors of omission and deploy a method free from both. But the realities do not permit this. Any method of inquiry that is operable in real life is caught up in the sort of trade-off illustrated in figure 1.

From such a standpoint, it becomes clear that scepticism purchases the avoidance of mistakes at an unacceptable price. After all, no method of inquiry, no cognitive process or procedure that we can operate in this imperfect world, can be altogether failure free and totally secure against error of every description. Any workable screening process will let some goats in among the sheep. With our cognitive mechanisms, as with machines of nay sort, perfection is unattainable; the prospect of malfunction can never be eliminated, and certainly not at any acceptable price. Of course, we could always add more elaborate safeguarding devices. (We could make automobiles so laden with safety devices that they would become as large, expensive, and cumbersome as busses.) But that defeats the balance of our purposes. A further series of checks and balances prolonging our inquiries by a week (or a decade) might avert certain mistakes. But for each mistake avoided, we would lose much information. Safety engineering in inquiry is like safety engineering in life. There must be proper balance between costs and benefits. If accident avoidance were all that mattered, we could take our mechanical technology back to the stone age, and our cognitive technology as well.

The sceptics' insistence on safety at any price is simply unrealistic, and it is so on the essentially economic basis of a sensible balance of costs and benefits. Risk of error is worth running because it is unavoidable in the context of the cognitive project of rational inquiry. Here as elsewhere, the situation is simply one of nothing ventured, nothing gained. Since Greek antiquity, various philosophers have answered our present question, Why accept anything at all?, by taking the line that man is a rational animal. Qua animal, he must act, since his very survival depends upon action. But qua rational being, he cannot act availingly, save insofar as his actions are guided by his beliefs, by what he accepts. This argument has been revived in modern times by a succession of pragmatically minded thinkers, from David Hume to William James.

The contrast line of reasoning to this position is not, If you want to act effectively, then you must accept something. Rather, its line is, If you want to enter into the cognitive enterprise, that is, if you wish to be in a position to secure information about the world and to achieve a cognitive orientation within it, then you must be prepared to accept something. Both approaches take a stance that is not categorical and unconditional, but rather hypothetical and conditional. But in the classically pragmatic case, the focus is upon the requisites for effective action, while our present, cognitively oriented approach focuses upon the requisites for rational inquiry. The one approach is purely practical, the other also theoretical. On the present perspective, then, it is the negativism of automatically frustrating our basic cognitive aims (no matter how much the sceptic himself may be willing to turn his back upon them) that constitutes the salient theoretical impediment to scepticism in the eyes of most sensible people.

Sceptics throughout the ages have faced the objection that on their principles people cannot guide their activities by knowledge—that we cannot base our actions on the recognition that fire burns or that eating assuages hunger. David Hume put this point as follows: "But a Pyrrhonian . . . must acknowledge, if he will acknowledge anything, that all human life must perish, were his principles universally and steadily to prevail. All discourse, all action would immediately cease; and men remain in a total lethargy, till the necessities of nature, unsatisfied, put an end to their miserable existence."[9] To this sort of charge, the ancient sceptics always replied that action need not be based on knowledge at all. They insisted on the sufficiency of noncognitive guides for action, of practical motives for choosing to act one way rather than another. The conduct of life can be governed not by knowledge but by appearances: "we neither affirm nor deny . . . but we yield to those things which move us emotionally and drive us compulsorily to assent."[10]

As Sextus Empiricus himself insisted time and again, the springs of action are desire and aversion—seeking and avoiding—and these can operate without the intervention of any sort of credence, without our subscribing to a doctrine of any kind or endorsing any actual thésis to the effect that this or that *is really the case*. Life without knowledge, reason, or belief is certainly not in principle impossible: animals, for example, manage very well. Or again, a somewhat less radical strategy is available, one that countenances acceptance (and belief), but only on a wholly unreasoned basis (e.g., instinct, constraint by the appearances, etc). The sceptic can accordingly hold—and act on—all those beliefs that people ordinarily adopt, with only this difference: that he regards them as reflecting mere appearances and denies that the holding of these beliefs is rationally justified.

Hume's objection hits wide of the mark: scepticism need not immobilize action. However, it does destroy the prospect of *rational* action. Ludwig Wittgenstein wrote: "The squirrel does not infer by induction that it is going to need stores next winter as well. And no more do we need a law of induction to *justify (rechtfertigen)* our actions or our predictions."[11] This gets the matter exactly wrong. Had Wittgenstein written *perform* (or *carry out*) in place of *justify*, all would have been well. But once *justification (rechfertigung)* is brought upon the stage, induction or some functional equivalent for establishing the need is just exactly what does become necessary. Saying that someone is being rational commits us to holding that this person is in a position to rationalize what he says and what he does—to justify it and exhibit it as the sensible thing to do. The rational use of a technique inevitably requires a great deal of factual backing from our knowledge—

our purported knowledge—regarding how things work in the world. The rational man as such requires the availability of cogent reasons for action, and only with the abandonment of a rigoristic all-out scepticism can such reasons be obtained.

Still (so some sceptics argue) we need not go so far as to claim knowledge; reasonable belief will do. But this is a shuffling evasion. A rose by any other name is still a rose, and a claim that such-and-such provides the appropriate answer to a question about the world remains an informative commitment regardless of the words we use to endorse it. Both its informativeness and its riskness remain in place.

The principal point at issue is important and far-reaching. For various sorts of creatures, the noncognitive guide to action may well suffice. Action can be grounded by instinct without belief, so that the cognitive route to action is bypassed. But this is not possible for homo sapiens, for creatures constituted as we are. We humans are so designed that a suitable cognitive orientation to the world is to all intents and purposes a physical need on our part—of our very nature.

And this matter of cognitive risk taking has other important cognitive ramifications as well. Scientific inquiry (research) is an atypical search process, because we do not really know exactly what we are looking for. There is generally no prospect of saying in advance just what it is that we may eventually find. To be sure, the community as a whole will in all human probability achieve important results. But any individual engaged on a particular investigation faces the risk of total failure. Individual researchers have to proceed in the face of uncertainty, subject to the risk of inadequate returns on their efforts. One may come up empty-handed or, even should one discover something important, may be beaten there by someone else. The risk of failure inherent in scientific work explains both why the rewards of success are great in communal and public recognition (for the encouragement of individual effort) and why they are not correspondingly great monetarily (to enable the maintenance of an entire community). A Nobel laureate may be a thousand times better known and esteemed than an obscure worker but will be paid only two or three times as much. (Fortunately, the value structure of scientists is such that that is generally sufficient.)

Even with great incentives, however, scientific creativity demands a great deal of entrepreneurial spirit and risk acceptance. This is perhaps one of the reasons why Japan's contributions to creative science have not matched the potential of her human and material resources. Japan nowadays (1987) has half the population and almost half the gross national product of the United States, and yet the Japanese have won only five Nobel Prizes in the sciences, compared to almost 150 for the United States. Japanese society fosters an ethics that is conformist rather than individualistic, cooperative rather than competitive, and safety oriented rather than risk oriented (rather than accepting the possibility of failure and its accompanying loss of face). Japanese policy makers are rightly concerned to overhaul the school system so as to make it less committed to rote learning and more conducive to the encouragement of creativity. But perhaps the problem is more deep rooted, grounded in the lack of individualistic enterpreneurship and personal risk taking inherent in a cultural aversion to possible failure even given the prospect of great potential rewards.

AGAINST SCEPTICISM

Back to scepticism, however. In many real-life situations we are not so totally and unqualifiedly confident of our beliefs that we are willing "to bet the family farm" on them. These noncategorical beliefs are beliefs alright; we accept them as true, and not just as likely or plausible. But our acceptance is a hesitant one. (And note the crucial difference between actually accepting something as true, however hesitantly, and viewing it as *probably* true). Now, when it comes to the implementation of such guarded beliefs and unenthusiastic acceptance in situations that will exact real costs if we prove to be wrong, we would certainly want to hedge our bets. Consider the situation depicted in table 1. In a

TABLE 1
A Problem of Choice Between
Belief Implementation and Inaction

	Resulting Payoffs	
Eventuations	If I Act on My Belief	If I Play Safe
1. If matters are as I believe (probability p).	X	O
2. If matters are not as I believe (probability $1-p$).	x	O

situation of this sort, an expected value calculation along the standard lines would engender the following result:

$EV(act) = pX + (1-p)x = p(X-x) + x.$

$EV(play\ safe) = 0.$

On this basis, it follows that:

$EV(act) \rangle EV(play\ save)$ iff $p \rangle \dfrac{x}{x-X}.$

Thus on the standard principles of decision-theoretic rationality, it transpires that only if p is sufficiently large relative to the balance of the potential costs and benefits at issue will acting on the belief at issue be appropriate and advisable. Action on a (non-categorical) belief is not automatically validated by its status as such but hinges crucially on the detailed circumstances of the situation.

Accordingly, if all that a sceptic wanted was to remind us of the important fact that we should not automatically act on all our beliefs but should be mindful of the potential hazards involved—that we should not regard our beliefs as sacrosanct fixities—then his strictures would be altogether sensible and appropriate. But philosophical sceptics are much too radical for this cautious line, insisting that we should abstain altogether from belief and accept nothing whatsoever. And at this point, scepticism leaves mere prudent caution behind and slips into an unduly distrustful paranoia.

To be sure, such an argument against scepticism is

an essentially practical one. It does not establish the internal inconsistency or theoretical untenability of a sceptical position. Rather, it shows that the price we would pay in adopting the sceptical position is so high as to outweigh any real benefit that could possibly accrue. The basic demand for information and understanding presses in upon us and we must do—and are thus pragmatically justified in doing—whatever it takes to get it satisfied. The ultimate flaw of scepticism is that it is simply not cost effective in the satisfaction of our cognitive needs and wants. For—to reemphasize an earlier point—cognitive deprivation is as debilitating for us as dietary deprivation.

Such argumentation will not, of course, dislodge the committed sceptic from his stance of accepting no theses at all. How could it do so? Any mere *argumentation* is bound to be unavailing against the sceptic, because he will always simply reject the premises needed for the proposed argument. (This, after all, is wholly in character with the nature of his position.) All that argumentation can do is to forestall scepticism by showing the incompatibility of scepticism with sensible approaches sensibly accepted. But while an essentially economic refutation based on the balance of costs and benefits will not dislodge someone from a sceptical position in which he is already entrenched, it should discourage those who have not already taken this position from doing so in the first place.

The crucial fact is that inquiry, like virtually all other human endeavors, is not a cost-free enterprise. The process of getting plausible answers to our questions also involves costs and risks. Whether these costs and risks are worth incurring depends on our valuation of the potential benefit to be gained. And unlike the committed sceptic, most of us deem the value of information about the world we live in to be a benefit of immense value—something that is well worth substantial risks.

Philosophical epistemologists tend to view knowledge acquisition in purely theoretical terms and to abstract it from the crass business of exerting effort, expending resources, and running risks. But this is simply unrealistic. Even the purest and most theoretical of inquiries has its practical and mundanely economic dimension.

TWO

The Economics of Trust and Cooperation

SYNOPSIS

(1) Knowledge is power. But the hoarding of knowledge—monopolization, secretiveness, collaboration avoidance—is generally counterproductive. (2) In anything like ordinary circumstances, mutual aid in the development and handling of information is highly cost effective. (3) The way in which people build up epistemic credibility in cognitive contexts is structurally the same as that in which they build up financial credit in economic contexts. (4) Considerations of cost effectiveness—of economic rationality, in short—operate to ensure that any group of rational inquirers will in the end become a community of sorts, bound together by a shared practice of trust and cooperation.

THE COST EFFECTIVENESS OF SHARING AND COOPERATING IN INFORMATION ACQUISITION AND MANAGEMENT

In many ways, knowledge is power. Its possession facilitates impact and influence in the management of public affairs. It enables those who have "inside" information to make a killing in the marketplace. It opens doors to the corridors of power in corporations. It maintains experts in the style to which the present century has accustomed them, and assures that the U.S. government supports more think tanks than aircraft carriers.

Since information is power, there is a constant temptation to monopolize it. But information monopolies, however advantageous for some few favorably circumstanced beneficiaries, exact an awful price from the community as a whole. In this regard, sixteenth and seventeenth century science affords an admonitory object lesson. The secretiveness of investigators in those times—Isaac Newton preeminent among them—in matters of mathematics and astronomy assured that the development of natural science would be slow and difficult. In protecting the priority of their claims through secretiveness and mystification, the adepts of the day greatly impeded the development and dissemination of knowledge.[12] Only with the emergence of new means of information sharing to facilitate the diffusion of knowledge, such as academies and learned societies with their meetings and published proceedings, could modern science begin its sure and steady march. In particular, the open scientific literature can be seen as an effective and productive system for the authentication and protection of the stake of the creative scientist in the "intellectual property" created by his innovative efforts.[13]

While there are, of course, exceptions, in most circumstances of ordinary life and above all in natural science it pays all concerned to share information. From an economic point of view, we confront the classic format of a cooperation-inviting situation, where the resultant gain in productivity creates a surplus in which all can share to their own benefit. The evident advantages for the scientific community and its members of creating a system that provides inducement for people to promulgate their findings promptly, while at the same time imposing strong sanctions against cheating, falsification, and carelessness, militate powerfully in this direction. The open exchange of information in science benefits the work of the community, except possibly in those cases where secrecy can confer an economic advantage that can serve as a stimulatant to creative effort.

To be sure, secrecy survives even now in various nooks and crannies of the scientific enterprise, and good claims can sometimes be made on its behalf. For example, editors of scientific journals do not let authors of submitted papers know the identity of those who review their submissions. Or again, such journals do not publish lists of authors whose submissions have been declined for publication. These practices obviously serve the interests of the journal's effectiveness by helping to maintain pools of willing reviewers and submitters.

The fundamental principles are the same either way, however. Both the general policy of information sharing in science, and the specific ways in which particular practices standardly depart from it, have a perfectly plausible rationale in cost-benefit terms. Broadly economic considerations of cost and benefit play the determinative role throughout.

THE ADVANTAGES OF COOPERATION

Even people who do not much care to cooperate and collaborate with others are well advised in terms of their own interests to suppress this inclination. This point is brought home by considering the matter from the angle presented in table 2. By hypothesis, each of the parties involved prioritizes the situation where they are trusted by the other while they themselves need not reciprocate. And each sees as the worst case a situation where they themselves trust without being trusted. Each, however, is willing to trust to avert being mistrusted themselves. Relative to these suppositions, we arrive at the overall situation of the interaction matrix exhibited in table 3. (Here the entry 2/2—for example—indicates that in the particular case at issue the outcome ranks 2 for me and 2 for you, respectively.) In this condition of affairs, mutual trust is the best available option—the only plausible way to avert the communally unhappy result 3/3.[14] In this sort of situation, cooperative behavior is obviously the best policy. (We are, after all, going to end up acting alike since, owing to the symmetry of the situation, whatever constitutes a good reason for you to act in a certain way does so for me as well.)

It is easily seen that a sceptical presumption—one which rejects trust and maintains a distrustful stance toward the declarations of others—would confront us with an enormously complex (and economically infeasible) task for the project of interpersonal communication. For suppose that, instead of treating others on the basis of *innocent until proven guilty*, one were to

TABLE 2
A Preferential Overview of Trust Situations

I Trust You	You Trust Me	My Preference Ranking	Your Preference Ranking
+	+	2	2
+	−	4	1
−	+	1	4
−	−	3	3

TABLE 3
An Interaction Matrix for Trust Situations

	You Trust Me	You Do Not Trust Me
I trust you	2/2	4/1
I do not trust you	1/4	3/3

treat them on the lines of *not trustworthy until proven otherwise*. It is clear that such a procedure would be vastly less economic. For we would now have to go to all sorts of lengths in independent verification. The problems here are so formidable that we would obtain little if any informative benefits from the communicative contributions of others. When others tend to respond in kind to one's present cooperativeness or uncooperativeness, then no matter how small one deems the chances of their cooperation in the present case, one is nevertheless well advised to act cooperatively. As long as interagents react to cooperation with some tendency to reciprocation in future situations, cooperative behavior will yield long-run benefits.

From the angle of economy there are, accordingly, substantial advantages to collaboration in inquiry, particularly in scientific contexts. For the individual inquirer, it decreases the chances of coming up completely empty-handed (though at the price of having to share the credit of discovery). For the community, it augurs a more rational division of labor through greater efficiency by reducing the duplication of effort.

BUILDING UP TRUST: AN ECONOMIC APPROACH

The process through which mutual trust in matters of information development and management is built up among people cries out for explanation by means of an economic analogy that trades on the dual meaning of the idea of credit. For we proceed in cognitive matters in much the same way that banks proceed in financial matters. We extend credit to others, doing so at first to only a relatively modest extent. When and if they comport themselves in a manner that shows that this credit was well deserved and warranted, we proceed to give them more credit and extend their credit limit, as it were. By responding to trust in a responsible way, one improves one's credit rating in cognitive contexts much as in financial contexts. The same sort of mechanism is at work in both cases: recognition of credit worthiness engenders a reputation on which further credit can be based; earned credit is like money in the bank, well worth the measures needed for its maintenance and for preserving the good name that is now at stake.

And this situation obtains not just in the management of information in natural science but in many other settings as well, preeminently including the information we use in everyday-life situations. For example, we constantly rely upon experts in a plethora of situations, continually placing reliance on doctors, lawyers, architects, and other professionals. They too must so perform as to establish credit, not just as individuals but, even more crucially, for their profession as a whole.[15]

Much the same holds for other sources of information. The example of our senses is a particularly important case in point. Consider the contrast between our reaction to the data obtained in sight and dreams. Dreams, too, are impressive and significant seeming. Why then do we accept sight as a reliable cognitive source but not dreams—as people were initially minded to do? Surely not because of any such substantive advantages as vividness, expressiveness, or memorability. The predisposition to an interest in dreams is clearly attested by their prominence in myth and literature. Our confident reliance on sight is not a consequence of its intrinsic preferability but is preeminently a result of its success in building up credit in just the way we have been considering. We no longer base our conduct of affairs on dreams simply because it doesn't pay.

Again, a not dissimilar story holds for our information-generating technology—for telescopes, microscopes, computing machinery, and so on. We initially extend some credit because we simply must, since they are our only means for a close look at the moon, at microbes, and so on. But subsequently we increase their credit limit (after beginning with blind trust) because we eventually learn, with the wisdom of hindsight, that it was quite appropriate for us to proceed in this way in the first place. As we proceed, the course of experience indicates, retrospectively as it were, that we were justified in deeming them creditworthy.

To be sure, the risk of deception and error is present throughout our inquiries: our cognitive instruments, like all other instruments, are never failproof. Still, a general policy of judicious trust is eminently cost effective. In inquiring, we cannot investigate everything; we have to start somewhere and invest credence in something. But of course our trust need not be blind. Initially bestowed on a basis of mere hunch or inclination, it can eventually be tested, and can come to be justified with the wisdom of hindsight. And this process of testing can in due course put the comforting reassurance of retrospective validation at our disposal.

We know that various highly convenient principles of knowledge production are simply false:

- What seems to be, is.
- What people say is true.
- The simplest patterns that fit the data are actually correct.
- The most adequate currently available theory will work out.

We realize full well that such generalizations do not hold, however nice it would be if they did. Nevertheless we accept the theses at issue as principles of presumption. We follow the metarule: In the absence of concrete indications to the contrary, proceed as though such principles were true. Our standard cognitive practices incorporate a host of fundamental presumptions of initial credibility, in the absence of concrete evidence to the contrary:

- Believe in your own senses.
- Accept at face value the declarations of other people (in the absence of any counterindications and in the absence of any specific evidence undermining their generic trustworthiness).
- Trust in the reliability of established cognitive aids and instruments (telescopes, calculating machines, reference works, logarithmic tables, etc.) in the absence of any specific indications to the contrary.
- Accept the declarations of established experts and authorities within the area of their expertise (again, absent counterindications).

The justification of these presumptions is not the factual one of the substantive generalization, In proceeding in this way, you will come at correct information and will not fall into error. Rather, it is methodological justification. In proceeding in this way, you will efficiently foster the interests of the cognitive enterprise; the gains and benefits will, on the whole, outweigh the losses and costs.

Such principles of presumption characterize the way in which rational agents transact their cognitive business. Yet we adopt such practices not because we can somehow establish their validity, but because the cost-benefit advantage of adopting them is so substantial. The justification of trust in our senses, in our fellow inquirers, and in our cognitive mechanisms ultimately rests on considerations of economic rationality. And this sort of situation prevails in many other contexts. For example, the rationale of reputations for ability, as well as those for reliability, lies in the cost effectiveness of this resource in contexts of hiring, allocating one's reading time, and so on.[16]

It is clear that all such cognitive practices have a fundamentally economic rationale. They are all cost effective within the setting of the project of inquiry to which we stand committed by our place in the world's scheme of things. They are characteristics of the cheapest (most convenient) way for us to secure the data needed to resolve our cognitive problems—to secure answers to our questions about the world we live in. Accordingly, we can make ready sense of many of the established rules of information development and management on economic grounds. By and large, they prevail because this is maximally cost effective in comparison with the available alternatives.

To be sure, whenever we trust, matters can turn out badly. In being trustful, we take our chances (though of course initially in a cautious way). But one must always look to the other side of the coin as well. A play-safe policy of total security calls for not accepting anything, not trusting anyone. But then we are left altogether empty-handed. The quest for absolute security exacts a terrible price in terms of missed opportunities, forgone benefits, and lost chances. What recommends those inherently risky cognitive policies of credit extension and initial trust to us is not that they offer risk-free sure bets but that, relative to the alternatives, they offer a better balance of potential benefits over potential costs. It is the fundamentally *economic* rationality of such cognitive practices that is their ultimate surety and warrant.

A COMMUNITY OF INQUIRERS

Only through cooperation based on mutual trust can we address issues whose effective resolution makes demands that are too great for any one of us alone. In the development and management of information, people are constantly impelled toward a system of collaborative social practices—an operational code of incentives and sanctions that consolidates and supports collective solidarity and mutual support. In this division of labor, trust results from what is, to all intents and purposes, a custom consolidated compact to conduct their affairs in friendly collaboration.

If its cognitive needs and wants are strong enough, any group of mutually communicating, rational, dedicated inquirers is fated in the end to become a *community* of sorts, bound together by a shared practice of trust and cooperation, simply under the pressure of its economic advantage in the quest for knowledge.

However, this cooperative upshot need not ensue from a moral dedication to the good of others and care for their interests. It can emerge for reasons of prudential self-interest alone, because the relevant modes of mutually helpful behavior—sharing, candor, and trustworthiness—are all strongly in everyone's interest, enabling all members to draw benefit for their own purposes—the agent himself specifically included. Cooperation emerges in such a case not from morality but from self-interested considerations of economic advantage. In science, in particular, the advantages of epistemic values like candor, reliability, accuracy, and the like, are such that everyone's interests are well served by fostering adherence to the practices at issue.

The pursuit of knowledge in science can play a role akin to that of a pursuit of wealth in business transactions. The financial markets in stocks or commodities futures would self-destruct if the principle, *my word is my bond*, were abrogated, since no one would know whether a trade had actually been made. In just this way, too, the market in information would self-destruct if people's truthfulness could not be relied upon. Thus in both cases, unreliable people have to be frozen out and exiled from the community. In cognitive and economic contexts alike, the relevant community uses incentives and sanctions (artificially imposed costs and benefits) to put into place a system where people generally act in a trusting and trustworthy way. Such a system is based on processes of reciprocity that advantage virtually everyone.

Several recent studies illuminate the extent to which we actually depend upon others in our beliefs.[17] The experiments of Solomon Asch have dramatized people's tendency to conform to erroneous public judgments on matters where they would never make mistakes by themselves.[18] His subjects had only to specify which of three lines was closest in length to a given line. People made this judgment unerringly, except when they knew that all the others who were asked the same question concurred in giving a different answer.[19] Commenting on Asch's experiments, Sabini and Silver report: "All (or nearly all) subjects reacted with signs of tension and confusion. Roughly one-third of the judgments subjects made were in error. Nearly 80 percent of the subjects gave the obviously wrong answer on at least one trial. The perception that a few other people made an absurd judgment of a clear, unambiguous physical matter was a very troubling experience, sufficient to cause doubt, and in some cases conformity."[20] Such experiments actually reveal (in their own dishonest way) the extent to which people incline to trust others. A recent study of American juries arrived at very similar findings.[21] On examining more than 250 jury deliberations, the investigators found that in no case was a hung jury caused by a single dissenter. Unless someone who disagreed with the majority found support by at least two others, the dissenters relaxed their reservations and came around to the majority view. And the rationale for this sort of thing is validated by sound economic considerations. a trusting relationship reduces current interaction costs in return for past investments in its buildup. Knowing whom one can trust is worth a great deal. Outsiders coming as strangers into an established social framework generally have to pay for the benefit of learning which agents are trustworthy—and generally find this information well worth paying for.

Such considerations militate for a universally advantageous modus operandi, under whose aegis people can trust their fellows in a setting of communal cooperation. And the harsh measures used to uphold the integrity of science—the destruction of careers through ostracism from the community—are thus not devoid of rational justification. Cheating is worth eliminating at great cost, because its toleration endangers and undermines the fabric of mutual trust, in whose absence the whole enterprise of collaborative inquiry becomes infeasible. establishing and maintaining a community of inquirers united in common collaboration by suitable rewards and sanctions is a mode of operation that is highly cost effective. Individual probity and mutual helpfulness are virtues whose cultivation pays ample dividends for the community of inquirers. In these matters, the cold iron hand of individual and communal interest lies behind the velvet glove of etiquette and ethics. The commodity of information illustrates rather than contravenes the division of labor that results from Adam Smith's putative innate human "propensity to truck, barter, and exchange." The market in knowledge has pretty much the same nature and the same motivation as any other sort of market—it is a general-interest arrangement. Cooperation evolves because what is in the interests of most is, in most cases, in the interests of each.

As these deliberations indicate, our cognitive practices of trust and presumption are undergirded by a justificatory rationale whose nature is fundamentally economic. For what is at issue throughout is a system of procedure that assures for each participant the prospect of realizing the greatest benefit for the least cost. Our standard cognitive policies and procedures are sustained by a basis of economic rationality.

THREE

Economic Aspects of Communication

SYNOPSIS

(1) Communication is predicated on conceding and maintaining credibility. In this light, some of its basic principles and presumptions can be validated by fundamentally economic considerations. (2) Informative exchange based on reciprocity and trust is a mutual benefit process. (3) Many of the cardinal aspects of scientific communication can be explained on the basis of economic considerations. (4) The differences between the communicative situation in science and ordinary life can also be accounted for in this way.

COMMUNICATION REQUIRES CONCEDING AND MAINTAINING CREDIBILITY

In cognitive contexts, communication is the process of conveying information from one person to another—or at any rate endeavoring to do so. Its ways and means are governed by economic considerations to an extent that is not generally appreciated. The exchange of information is clearly a mutual benefit process. It is far easier, cheaper, and more convenient for people to get information by sharing than by themselves having to undertake the often laborious inquiries and researches needed to develop it *de novo*.

To derive benefit from the declarations of others, we must (1) listen to them, pay them heed; (2) interpret (decode) what they say; (3) extend them cognitive credit. To be sure, none of these steps is cost free. Each exacts from us an outlay of resources in point of time, effort, attention, and risk of error. All procedures for the acquisition of information—listening, watching, reading, and so on—involve expenditures of some sort. (No school student is ever wholly oblivious to the fact that learning can be painful.) And whether this outlay is warranted depends on the correlative advantages—preeminently including the cognitive benefits of acquired information.

Suppose I tell you, The cat is on the mat. What information do you now actually have? Is it (1) The cat *is* on the mat? Or is it (2) Rescher *thinks* (believes) that the cat is on the mat? Or is it merely (3) Rescher *says* that the cat is on the mat? In the circumstances under hypothesis, only the last item is wholly unproblematic. And it is clear that one cannot manage to get from (3) to (2) unless one adds something like (4), When Rescher says something (in a serious tone of voice) he generally believes it. Moreover, you certainly cannot get from (2) to (1) unless you credit me with veracity and trustworthiness and impute to me a penchant for truth—that is, unless you also accept (5) When Rescher believes something to be so (in such matters as cats and mats, at any rate) he is generally right.

The entire process of communication—of deriving substantive information from the declarations of others—involves trust. But what validates this? To answer this question it is best to look at the issue in economic perspective.

A communicating community is a sort of marketplace with offerers and takers, sellers and buyers. In accepting the declarations of others at their informative face value, we extend them credit, as it were. The prospect of informative communication is predicated on such principles as (i) Concede a presumption of veracity to the assertions of others, at any rate until such time as they prove themselves unworthy of credit; and (ii) In communicative contexts, regard others as candid truthful, accurate, and the like, until proven otherwise. The rationale of such principles of epistemic procedure is largely or wholly economic. for here, as elsewhere, it is ultimately on the basis of considerations of cost effectiveness that we decide how much credit to extend.

After all, why do we credit people with communicative capacity—with the power to provide information? Note that the purported fact, *When he utters "the cat is on the mat," he is engaged in asserting that the cat is on the mat*, represents a belief of ours, or at any rate a supposition on our part. We make this supposition initially in desperation, as it were, because it provides the only feasible way for us to derive any benefit from the content of someone's assertions, but ultimately because we eventually accumulate evidence that indicates (with the wisdom of hindsight) that this supposition was well advised (warranted).

In extracting information from the declarations of

others we rely on a whole host of working assumptions: (1) People mean to say what they apparently say (what we take them to be saying). (2) People believe what they say. (3) People have good grounds for their beliefs (i.e., there are such grounds and they have cognizance of them). Essentially the same justification obtains throughout: unless we enter into such communicative presumptions, we deprive ourselves of any chance to extract information from the declarations of others. On this basis, we are rationally well advised (for example) to treat their declarations as epistemically innocent until proven otherwise, exactly because this is the most cost effective thing to do. Our communicative procedures are motivated—and justified—by the essentially profit-seeking objective of extracting the maximum benefit from our information-oriented interactions.

Communication is accordingly predicated on conceding and maintaining credibility. Communication too is a commercial system of sorts. Credit is extended, drawn upon, and enlarged. And with communicative and financial credit alike, one could not build up credit (prove oneself creditworthy) unless given *some* credit by somebody in the first place. For credit to be obtainable at all, there has to be an initial presumption that one is creditworthy. Clearly, such a presumption of innocent until proven guilty (i.e., fault free until shown to be otherwise) can be defeated; one can of course prove oneself to be unworthy of credit or credence. But initially it must be made.

The guiding principle here is that of cost-benefit calculation. The standard presumptions that underlie our communicative practices are emphatically *not* validatable as established facts. (For example, it is certainly *not* true that people say what they mean, save at the level of statistical generality.) But their justification becomes straightforward on economic grounds, as practices that represent the most efficient and economical way to get the job done. For if we do not concede *some* credit to the declarations of others, then we lose any and all chance to derive informative profit from them, thus denying ourselves the benefit of a potentially useful resource. For the course of experience would soon teach us that, even where strangers outside the family circle are concerned, the benefits of trust, of credibility concession, generally overbalance the risks involved.

In sum, we hold to the policy of believing what we are told in the absence of case-specific counterindications, because it is in our interest to do so by being highly cost effective vis-à-vis our informational aims and purposes. If playing safe were all that mattered, we would, of course, suspend judgment indefinitely. But it is simply not in our interest to do so, since safety is not all. We adopt the policy of credence in the first instance because it is the most promising avenue toward our goals, and then persist in it because we subsequently find, not that it is unfailingly successful, but that it is highly cost effective.[22]

COMMUNICATION AS A GENERAL BENEFIT ENTERPRISE

Information exchange based on principles of cooperation is a mutual benefit process, for everyone is advantaged by adopting a system of operation that maintains the best available balance of costs and benefits in this matter of creating a communally usable pool of information.

Contrast two hypothetical societies of communicators, the Liars and the Deceivers. The Liars generally say the opposite of what they think: "Rotten day today," they say when the weather is beautiful, and vice versa. The Deceivers, however, do not behave so reliably: they mix putative truth and falsity more or less randomly. The Liars can communicate with us and with each other perfectly well. Once one has got the trick, one knows exactly how things stand in discussions with them. But the Deceivers are something else again. One never knows where one stands with them. And worse yet, they too have no idea where they stand with one another. Indeed, they could never even begin to communicate. Even if an initial generation of Deceivers came equipped with a ready-made language (say, because they began as normal communicators and then turned en masse into Deceivers at some point), the fact remains that they could never teach language to their offspring. "That's a lion," the parents observe to Junior one day, pointing to a dog, and "That's a cat," the next time, and "That's an elephant," the time after that. Poor Junior would never manage to catch on.

Contrast now two other communities: The Trusters and the Distrusters. The Trusters operate on the principle, be candid yourself, and also accept what other people say as truthful—at any rate in the absence of counterindications. The Distrusters operate on the principle, Be deceitful yourself, and look on the assertions of others in the same light—as ventures in deceitfulness: even when people are ostensibly being truthful, they are only trying to lure you into a false sense of security. It is clear once again that the policy of the Distrusters is totally destructive of communication. If exchange of information for the enhancement of our knowledge is the aim of the enterprise, then the diffusion of distrust is utterly counterproductive. To be truthful, to support the proper use of language and refrain from undermining its general operation is a policy of the greatest general utility—however beneficial occasional lapses may seem to be.

Not only is the maintenance of credibility as asset in communication, but some degree of it is in fact a necessary condition for the viability of the whole project. The precept, *Protect your credibility; do not play fast and loose with the ground rules, but safeguard your place in the community of communicators,* is basic to the communicative enterprise.

From the sender's point of view, putting forth a message costs time, effort, energy, and the like. The rational agent will incur such costs only with a view to benefits—some sort of reward (if only in the respect or gratitude of others) or reciprocity, with a view to a quid pro quo. This point is simple but of far-reaching import. Given our need for information to orient us in the world (on both pure and practical grounds), the value of creating a community of communicators is enormous. We are rationally well advised to extend ourselves to keep the channels of communication to our fellows open, and it is well worth expending much for the realization of this end.

And the same sort of story holds for the receivers' point of view as well. They too must expend resources on taking in, processing, and storing messages. Clearly, a rational hearer would be prepared to undertake this expenditure only if there were a reasonable expectation of drawing profit from it, be it by way of information obtained or pecuniary reward—an expectation which, in general, is amply warranted.

In this regard, it is useful to consider H. P. Grice's influential interpretation of communicative practice.[23] Grice's analysis focusses on the intentions of speakers: the speaker aims to induce his auditor to accept (believe) something by getting him to recognize that this is what the speaker is seeking to do. This is correct enough as far as it goes. But Grice's speaker's perspective slights the fact that an equal burden is borne by the hearer in communication situations. To begin with, the auditor will not even enter into communicative contact unless persuaded by the prospect of some benefit. If he did not think that the speaker was trying to convey something to him—if, for example, he regards the movement of his lips as we regard the twitching of a sleeping dog's legs—he would give little heed. Moreover, the hearer has to use whatever cues and clues come to hand to construct an interpretation of what is (putatively) being said. The speaker's intentions notwithstanding, the only information that is ever actually *conveyed* is what the hearer accepts. And this is, in general, a matter of the simplest, the most cost-effective construction that he can place upon the speaker's declarations relative to what he takes to be the questions at issue, using whatever data that come to hand, be they tacitly behavioral or overtly expressed.

The speaker, of course, benefits in communicative

situations insofar as the information conveyed influences the hearer's present and future actions in directions the speaker sees as desirable. And failures of transmission are generally of detriment no less to those who convey information than to those who receive it. In informative communication, the hearer—who is a voluntary party to the transaction—wants to extract the maximum benefit (in terms of answers to his questions) from the speaker's assertions—subject to protecting the speaker's credibility insofar as possible. This is a matter of a common interest between the two parties, for once the speaker's reliability is lost (in the hearer's opinion), then his utility as a source of information is utterly destroyed. Hence speaker and hearer have a joint interest in protecting the speaker's credit.

Consider just one example. Suppose you and I communicate (in the informative mode). In the first instance, for the economic reasons already canvassed, I extend you benefit of doubt and take your various declarations at face value. I enter into the discussion with an established view of the world already in place, and I use your declaration to extend and amplify it. You say, "The cat is on the mat," and I proceed to readjust my world picture to place the cat on the mat. Your information supplements my information base and enlarges the details of my world picture.

But the matter is very different when you present discordant information. If, for example, you say, "A unicorn is in the forest," then I shall make no change on my world picture save in relation to you and your state of mind. I endorse: "He is under the (mis)impression that there is a unicorn in the forest." That is, when the result of integrating the substance of your declarations into my world picture becomes too discrepant (when too great a cognitive cost would ensue from my endorsement of your claims, because I must abandon too much to accommodate them), then instead of supplementing my register of endorsed propositions with your claim p, I merely add, "He has the (mis)understanding that p." So when you say, "The present king of France is bald," then instead of taking your declaration at face value, I endorse: "He has the mistaken idea that there is currently a king of France who is bald." And here no ontological novelty is introduced at all, because the claim I endorse is one about *you*, and you are already featured on my ontological agenda. The need for making special ontological provision for nonexistents simply does not arise. From the ontological point of view, we can deal with claims regarding nonentities wholly in terms of existents, namely in terms of the beliefs, (mis)impressions, assumptions, suppositions, and the like, of perfectly real people. On economic grounds, a new ontological category of nonexistent entities is quite dispensable. We can make do with real people and what they can know, think, and suppose. The manner in which nominalistic philosophers stress ontological economy in the context of talk about nonexistent possibles can accordingly also be understood in terms of quasi-economic costs and benefits.

In general, when one endeavors to convey information to someone, various sorts of unpleasant reactions can occur—as table 4 shows.

Throughout, effective communication is a matter of

Economic Aspects of Communication

TABLE 4
Communicative Negativities and Their Causes

Negativity	Sample Causative Etiology
One is disbelieved.	One speaks falsely too often (out of heedlessness or out of deceptiveness).
One is misunderstood.	One uses careless or inadequate formulations.
One is tuned out.	One speaks off the point (digresses), or speaks at undue length (even if it is to the point).

maintaining proper cost-benefit coordination. It is governed by such principles as

- Be sufficiently cautious in your claims to protect your credibility, but do not say so little that people dismiss you as a useless source.
- Formulate your statements fully and carefully enough so as to avoid misunderstanding but not with so much detail and precision as to weary your auditors and get tuned out.
- Make your message long (explicit, detailed, etc.) enough to convey your points but short enough to avert wasting everyone's time, effort, and patience.
- Be sufficiently redundant that an auditor who is not intensely attentive can still get the point, but not so redundant as to bore or annoy or insult your auditors.

- Keep to the point but not so narrowly that your message is impoverished by lack of context.

All of these principles are fundamentally economic principles of balance; they all turn on finding a point after which the benefit of further gain in information falls below the cost demanded for its process of acquisition.

ASPECTS OF SCIENTIFIC COMMUNICATION

The situation of science clearly illustrates the preceding strictures about the cost-benefit advantages of trust and cooperation. Every scientist depends on others not only for his training but also for the starting point of information, without which he could not begin his inquiries. Cooperation is essential in testing and corroborating scientific claims and in setting up and implementing the standards by which the distinction between genuine science and crackpot speculation can be maintained.

But cost-benefit considerations also enter in at many other points to explain the handling of scientific information. The same economies of scale, resulting from the efficiencies of mutual access, that draw people together in cities draw scholars and scientists together in universities, research centers, institutes, academies, and professional associations. Had we not inherited such collectivities, we would now have to reinvent them. And much the same holds good for professional conferences, journals, preprint exchange networks, and the like.

With the desktop computer or word processor, the information revolution launched by Gutenberg reached its apogee. Just as the printed book is a splendidly effective device for realizing economies of scale in the *communication* of information, so the word processor is a comparably splendid device for the processing of information. The book and the computer match and complement each other perfectly, the one consumer oriented the other producer oriented.

The free exchange of the scientific literature reflects the status of the enterprise as a mutual assistance society. Even if individual credit for one's own contributions were not forthcoming through professional recognition, it would still be well worthwhile—insofar as intellectual curiosity is a significant motive for them—for individuals to enter into such an arrangement, simply in the interests of fostering the conditions of their own work.

It warrants note, moreover, that even apart from such considerations of rewards and incentives, powerful considerations of economic cost effectiveness militate against anonymity in scientific and scholarly publications. Identifying authors is advantageous through enabling readers to make some initial discrimination between insignificant material and the presumably valid work of people who, having gained good reputations through competent work, have something at risk. Blind publication would have the substantial disadvantage of a loss of useful information. (Contemplate a scientific journal of anonymous results!) The practice of publication and citation in science—of claiming and giving credit—is so effective a way of harmonizing private aims with public benefits as to seem the product of an idealized social contract drawn up by the scientific community.[24]

It is illuminating in this regard to consider the reward system of science. Why do scientists in evaluating people's contributions accord such great value to priority? After all, as long as the work was done independently and accomplished within the same state of the art, the achievement is surely just as great. Why make credit for scientific discovery a winner-takes-all process, as in political elections? The answer lies in part in the circumstance that this provides a maximum incentive to creative effort. Moreover, the interests of the community as a whole require avoiding a duplication of effort. The originality-promoting principle, that once done it's a dead issue, gives powerful assurance that people will not work in dried-up areas. The reward system of science is by and large designed to promote aims and objectives of the enterprise in the most efficient way. It is the product of an evolutionary emergence of processes that are cost effective in serving the long-range and overall interests of the community by means of addressing the short-range and individual interests of its constituent members.

The publication system of science also exhibits interesting economic features. For example, the scientific literature consists of modes of publication that vary in length substantially—indeed, in a roughly exponential way:

- A note or letter or abstract (1 page): 1.
- A short article (4–15 pages): 5.

- A long article (20–40 pages): 25.
- A monograph (100–50 pages): 125.
- A book (200–800 pages): 625.
- A multivolume book series (2,000–10,000 pages): 3,125.

In the scientific literature there is thus a general coordination between problem size and text size, for the different sorts of publications closely correspond to issues of different sorts and sizes:

- For a problem; a report (of observation or experiment) or a note.
- For a problem cluster; a paper or a short article.
- For a problem area; a review article or a full-scale article.
- For a subfield; an at-length survey or a monograph.
- For a field; a treatise or a book.
- For a discipline; an encyclopedia, a handbook, or a multivolume series.

Increasing length is thus proportionate to the increasing size of the substantive issues dealt with. In scientific publication, there tends to be a cost-benefit-governed coordination between problem size and text size, a coordination that proportions the physical space (and thus the reading time and expense of production and acquisition) devoted to an issue in accordance with its substantive significance. The fundamental economic principle of allocating space in line with inherent importance is thus operative.

SCIENTIFIC VERSUS ORDINARY-LIFE COMMUNICATION

It is useful to bear in mind that different priorities obtain in different contexts of communication. In everyday-life communication where we are deeply concerned to protect our credibility, we value security over informativeness. Hence looseness and imprecision are perfectly acceptable. On the other hand, in science, we value generality and precision over security. After all, natural science is not content with theses like, *On the whole, larger objects are heavier*, or *Most things made predominantly of lead generally melt at temperatures around 330 degrees Celsius*. In science, we seek exactness and precision: we want to know how all objects of exactly this or that sort always behave. Generality, precision, detail are at a premium, and so in scientific discourse we prioritize these factors in a way that makes our scientific theories vulnerable. (The half-life of theories in frontier physics is relatively short.)

It is of the nature of natural science at the research frontier that it aims to characterize nature's processes exactly and to describe how they operate always and everywhere, in full generality and precise detail. Technical science forswears the looseness of vague generality or analogy or approximation. It has no use for qualifiers such as usually, normally, or typically; universality and exactness are its touchstones. Science, accordingly, declares not merely that roughly such-and-such generally occurs in certain sorts of circumstances, but exactly what happens in exactly what circumstances. In science we always aim at the maximum of universality, precision, and exactness. The law claims of science involve no hedging, no fuzziness, no incompleteness, and no exceptions; they are strict: precise, wholly explicit, exceptionless, and unshaded. In making the scientific assertion, "The melting point of lead is 327.7 degrees Celsius," we mean to assert that *all* pieces of (pure) lead will unfailingly melt at *exactly* this temperature. We certainly do not mean to assert that *most* pieces of (pure) lead will *probably* melt at *somewhere around* this temperature. (And in this regard, there would be a potential problem, should it turn out, for example, that there is no melting *point* at all and that what is actually at issue is the center of a statistical distribution.) And this commitment to generality and detailed precision renders the claims of science highly vulnerable. We realize that none of the hard claims of present-day frontier natural science will move down the corridors of time untouched. Fragility is the price that we pay in science for the sake of generality and precision.

Increased confidence in the correctness of our estimates can always be purchased at the price of decreased accuracy. We *estimate* the height of the tree at around 25 feet. We are *quite sure* that the tree is 25 ± 5 feet. We are *virtually certain* that its height is 25 ± 10 feet. But we are *completely and absolutely sure* that its height is between 1 inch and 100 yards. Of this we are completely sure, in the sense that we deem it absolutely certain, certain beyond the shadow of a doubt, as certain as we can be of anything in the world, so sure that we would be willing to stake our life on it, and the like. With any sort of estimate, there is always a characteristic trade-off relationship between the evidential *security* of the estimate on the one hand (as determinable on the basis of its probability or degree of acceptability), and its contentual *definiteness* (exactness, detail, precision, etc.) on the other.

This relationship between security and definiteness is generally characterized by a curve of the general form of an equilateral hyperbola: $s \times d = c$ (c = constant). (See figure 2.) The increased vulnerability and diminished security of our claims is the undetachable other side of the coin of the pursuit of definiteness. Science operates in the lower right-hand sector of the figure. Its cultivation of informativeness (definiteness of information) entails the risk of error in science: its claims are subject to great insecurity. No doubt the progress of science makes it possible to decrease the value of c somewhat, but the fundamental trade-off relationship remains unavoidable. An information-theoretic uncertainty principle prevents our obtaining the sort of information we would ideally like.[25] The exactness of technical scientific claims makes them especially vulnerable, notwithstanding our most elaborate efforts at their testing and substantiation.

The situation in science accordingly differs markedly from that which prevails in everyday life. When we ordinarily assert that peaches are delicious, we are asserting something like, Most people will find the eating of suitably grown and duly matured peaches a rather pleasurable experience. Such a statement has all sorts of built-in safeguards, such as *more or less, in ordinary circumstance, by and large, normally, if all things are equal,* and so on. They are nothing like sci-

FIGURE 2
The Relationship Between Security and Definiteness

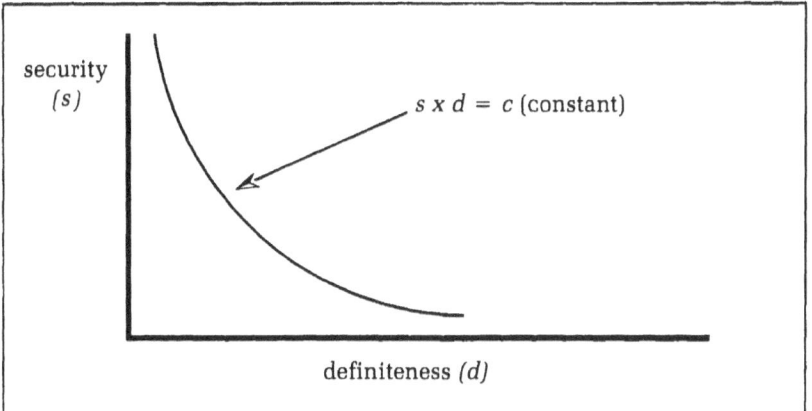

NOTE: The overall quality of the information provided by a claim hinges on combining its security with its definiteness. Given suitable ways of measuring security (s) and definiteness (d), the curve at issue can be supposed to be an equilateral hyperbola obtained with $s \times d$ as constant.

entific laws but mere rules of thumb, a matter of practical lore rather than scientific rigor. And this enables them to achieve great security, for there is safety in vagueness: a factual claim can always acquire security through inexactness. Take the claims, There are rocks in the world and Dogs can bark. It is virtually absurd to characterize such everyday-life generalizations as fallible. Their security lies in their indefiniteness or looseness; it is unrealistic and perverse to characterize such common-life claims as defeasible. They say so little that it is unthinkable that contentions such as these should be overthrown. And this accords smoothly with the needs of the situation. For ordinary-life communication is a practically oriented endeavor carried on in a social context. It stresses such maxims as: Aim for security, even at the price of definiteness; Protect your credibility; Avoid misleading people.

When, as in ordinary life, the preservation of credibility is paramount, one wants to formulate one's claims in as safe and secure a way as possible, and thus one resorts to vagueness and imprecision. On the other hand when, as in science, creativity and originality are paramount, then one would put one's claims in the most ambitious and surprising way, accepting the risks inherent in universality, precision, and the like. Plausibly enough, the appropriateness of an epistemic policy hinges upon the nature of the governing desideratum (credibility vs. creativity).

The aims of ordinary-life discourse are primarily *practical*, largely geared to social interaction and the coordination of human effort in communal enterprises that serve the common good. In this context, it is crucial that we aim at credibility and acceptance: that we establish and maintain a good reputation for reliability and trustworthiness. In the framework of common-life discourse, we thus take our stance at a point far removed from that of science. Very different probative orientations prevail in the two areas. In everyday contexts, our approach is one of situational satisficing: we stop at the first level of sophistication and complexity that suffices for our present needs. In science, however, our objectives are primarily theoretical and governed by the aims of disinterested inquiry. Hence the claims of informativeness—of generality, exactness, and precision—are paramount.

In science, we accept greater risks willingly because we ask much more of the project. We deliberately court risk by aiming at maximal definiteness and thus at maximal informativeness and testability. Aristotle's view that terrestrial science deals with what happens ordinarily and in the normal course of things has long ago been left by the wayside. The theories of modern natural science has little interest in what happens generally or by and large; they seek to transact their explanatory business in terms of strict universality, in terms of what happens always and everywhere and in all kinds of circumstances. And in consequence we have no choice but to acknowledge the vulnerability of our scientific statements, subject to the operation of the security-definiteness trade-off.

In ordinary life, we operate at the upper left-hand side of the figure 2 curve. The situation contrasts sharply with that of science, whose objectives are largely theoretical, and where the name of the game is rigorous understanding on a basis of unrestricted universality and extreme precision. The cost-benefit situation of the two domains is drastically different.

One recent student of scientific procedure maintains: "It [the modus operandi of science] is not an automatic consequence of the study of the natural world but it is a component of the code of behavior of the leading men. . . . Its presence is not guaranteed . . . but depends on accidents of history, and on the social and cultural environment in which the activity is coordinated."[26] But (sexism aside) this view offers a very problematic description of the situation. Whether or not (and to what extent) people dedicate themselves to a certain project (medicine, science, chess playing) is indeed a social matter. But once they do so—once they seriously and dedicatedly take such a project and its inherent teleology—then the inner mechanisms of the task situation constrain them toward the cost-effective means of the project-coordinated ends. There is then nothing accidental about their behavioral modes, for if they pursue the venture with intelligence and dedication, then rational selection will in the course of time inexorably bend their ways or proceeding into shapes that can be accounted for in terms of economic considerations.

FOUR

Importance and Economic Rationality

SYNOPSIS

(1) Importance is a key factor in the economics of cognition. For what is important is by virtue of this very fact more deserving of attention and effort than what is not. Importance is always comparative, a matter of the relative share of resources due to one item in comparison with others in the overall scheme of things. (2) Importance for understanding—cognitive importance—is the paradigmatically characteristic mode of importance. (3) The cognitive importance of things is not something that people somehow make up; it is objective. Unlike being interesting, being important does not lie in the eyes of the beholder. (4) The assessment of cognitive importance is a key issue for rationality in its economic concern for returns on resource expenditure.

THE ASSESSMENT OF IMPORTANCE

Economics calls for the assessment and balancing of costs and benefits. And importance is a key factor here, seeing that the rational management of our affairs calls for an allocation of effort (and other resources) in various directions in line with their actual importance.[27] For what is important is by virtue of this very fact more deserving of attention and effort than what is not. An item's importance is reflected in the volume of resources that its securing deserves, not in the volume of resources that its securing requires. Air to breathe is crucially important for us humans as a requisite for sustaining our very lives, even though we ordinarily need not expend much effort to obtain it (though when necessary we are prepared to go to considerable lengths to secure its availability, for example when travelling at great altitudes or depths).

Importance does not operate on egalitarian principles; in many or most cases, a few leading items usually predominate over the rest in this regard. Even where many factors make an essential contribution, comparatively few are generally of substantial importance. In this world, the distribution of significance, like that of frequency, tends to have a half-bell configuration, with some relatively few items taking the lion's share. (The probable reason for this is that, while various factors will generally be of some importance, nevertheless if all factors were equally crucial to an item, then Darwinian considerations would militate against its emergence and survival.)

A dictionary will define *importance* somewhat as follows: "having great significance, weight, consequence, or value." And it will go on to list such synonyms as *significance, essentiality, moment*, and such antonyms as *insignificance, inessentiality, negligibility*. The important things are clearly those that count and the unimportant ones those that don't. But how is the score to be kept?

Importance can be either conditional (instrumental) or absolute (intrinsic). Conditional importance is importance relative to contingent ends, goals that someone may have adopted—a knowledge of human physiology, say, for a physician; or physical agility for a tennis player. Absolute importance, on the other hand, relates to mandatory ends, goals that people should have (self-respect, say, or honesty, or concern for the well-being of their dependents). For while ends of course vary enormously with different people, there are some ends that people should have (self-respect, for example), which it is somehow reprehensible (foolish, perverse, crazy) for them to neglect, at any rate in the absence of more urgent factors impelling them in other directions at the moment. Accordingly, the importance of finding food is absolute (since survival is a universally appropriate goal), but the importance of knowing how to keep score at tennis is conditional in that it hinges on (say) one's idiosyncratic interest in playing or watching this game.

Importance turns on the extent to which the removal or diminution of a given item would undermine or diminish the prospects of realizing the aims, values, or functions at stake. The measurement of importance is thus a matter of determining the extent to which realization of relevant desiderata is aided by exploiting the matter whose importance is in question or is compromised by neglecting it. Accordingly, importance pivots on the idea of making a difference—of casting a large shadow across the particular issues in view. The pivotal question is: How large a penalty in reduced resources—lost time, money, understanding, or the like—would be entailed by the loss or neglect of the item whose importance is under consideration? And this is something that can change with changing circumstances. Obviously, what was once an important communication skill (knowing Latin) or an important military power (Sweden) may well cease to be so.

In assessing importance, it must be borne in mind that what matters is not just whether or not a certain factor is essential for a particular valued result, but the relative extent of its contribution. This is always comparative. The contribution to a result made by one factor cannot exceed 100 percent. Any one factor's gain in importance must be some other's loss. (However thin we slice it, there is just one pie.) The comparative share of the pie of a formerly important item may be diminished to the point of insignificance without any intrinsic change, as other factors become more prominent. (Think of how television has overshadowed the importance of radio as a mode of entertainment in the United States, even though there are more radio sets and radio programs than ever.)

COGNITIVE IMPORTANCE AS A TYPICAL CASE

In general, then, something is important where it makes a big difference for the realization of a function or end. Accordingly, something is of specifically cognitive importance to the extent to which it makes a substantial contribution to the realization of our cognitive purposes—knowing and understanding—in contrast, say, to its practical importance for our survival and achievement of chosen ends. The distinction between practical and cognitive importance is not hard and fast, however, because certain items of knowledge are also essential for life (knowing how to obtain food, for example).

Cognitive importance is the characteristic standard by which we assess the value of knowledge. It relates to what is important for understandinig—for the enlargement and improvement of the body of information at our disposal. It is perhaps the most characteristic and paradigmatic mode of importance. Moreover, it has many aspects and covers an immense area, for it not only bears upon directly cognitive items—ideas, books, theories, facts, experiments—but can also appertain derivatively to particular states of affairs or events, whose importance is obliquely dependent on their cognitive importance for understanding the march of events or the course of history.

The importance of a cognitive issue, or even an entire cognitive domain, is accordingly measurable by the volume of resources that its cultivation merits. Should it occupy 10 or 20 percent of a student's curriculum? Does an inquiry deserve a 5 or 10 percent share of the resources we dedicate to scientific pursuits? Throughout such contexts, we have in view an overall pie to be divided into pieces of different relative size; we face such questions as whether an avenue of research merits the investment of 6 or 16 percent of our overall budget of resources of time, attention, and money.

One recent writer suggested that the value of knowledge lies wholly in its utility for resolving the practical problems of human life, so that all knowledge is only instrumentally important for meeting our practical needs and wants.[28] But this line is very questionable. For knowledge also bears satisfactions of its own, seeing that it may, quite appropriately, be pursued for the sheer joy of understanding. The aforementioned author condemned such theoretical satisfaction as reflecting a "rationalistic neurosis."[29] But this is mere name calling. Inquiry, like artistic creativity, produces results whose contemplation can provide deep satisfaction. In the course of scientific progress, those who make important discoveries are for that reason more amply compensated by way of recognition and rewards. And this is only fitting and proper, given the nature of the enterprise. From classical antiquity onward, the Western cultural tradition has stressed benefits of the pursuit of knowledge with a fervor that is unlikely to be a matter of sheer self-delusion. Our very nature as *homo sapiens* is at stake in the enhancement of knowledge for knowledge's sake. If this is a neurosis, we would do well to make the most of it.

As noted above, cognitive importance can be either instrumental (conditional) or intrinsic (unconditional). Unconditional cognitive importance is intrinsic to our grasp of things in general—for securing understanding on a very broad and encompassing scale. By contrast, conditional cognitive importance relates to a more narrowly circumscribed subject or topic or particular technical interest. Different issues are involved. Topics highly important for knowing about the care and nurture of guinea fowl, for example, may play a very insignificant role in understanding the overall scheme of things. The crux is a matter of generality versus special interest.

Cognitive importance is enormously complex. Cities are sociologically important because they are so populous, desert environments because they are so sparsely populated. The illuminative bearing of extreme conditions endows them with particular importance. Importance thus exhibits polarity—opposite extremes may be equally powerful as import-indicative considerations. In the case of cognitive values, commonness and rarity, largeness and smallness, typicality and aberration can all provide bases for cognitive importance. The particular factors that determine cognitive importance are thus endlessly varied.

A particular inquiry may be important only in the context of others. If it is important to pinpoint an object located in any one of four thousand places, and each of a hundred investigators is assigned forty places to search, then it becomes critically important that each one of us should do our particular part of the inquiry correctly. Though in itself each bit has only small importance, division of labor may render the contribution of each piece an important facet of a larger mosaic.

Other things being equal, cognitive importance pivots on such parameters as (inherent) *significance, centrality, generality,* and *fertility.* Cognitive *significance* is a matter of serviceability for achieving a satisfactory cognitive orientation toward the world about us. It pivots on the value of knowledge for knowledge's sake; that is, for realizing the satisfactions of understanding, as such. *Centrality* is a matter of the linkage of one item to others: the more central, the more extensively interconnected with other items. The role of *generality* hinges on the fact that the broader and more inclusive the range and reach of an item, the more fully it extends our cognitive grasp. *Fertility* is something else again—a matter of an item's role in opening up vistas toward the understanding of new, heretofore unprobed issues.

The structure of the cognitive domain can be depicted in a kind of map akin to the displays used to illustrate the relative sizes of the cities in a country and the transport linkages among them. Significance reflects the comparative size of a given unit. Centrality is determined by the extent of its enmeshment in the overall network—by how many links this unit has with others. Generality and fertility are reflected in the volume of traffic moving along these links.

Is importance all-powerful in cognitive matters? Should we perhaps concentrate *all* of our cognitive energies for information development, storage, and retrieval on that which is, as best we can tell, the most important? This sensible idea faces formidable difficulties. The problem lies in the gap between apparent importance and real importance. Ample experience teaches that many of the things that currently seem unimportant will eventually come to appear otherwise. (Think of Einstein and the perihelion of Mercury; or Becquerel and his photographic plates.) In information development, storage, and retrieval, there is need for a broad-gauge, many-sided approach. In prospecting for petroleum, we must be prepared to drill many wells, because no one can say in advance of the event just where oil will be struck. The situation in inquiry is not dissimilar. More often than not, cognitive importance can be discerned only with the wisdom of hindsight, as its implications and ramifications become more apparent. Only with the theory of relativity did the anomalous variations of Mercury's perihelion become an important issue, and only with the rise of the computer did binary coding become an important mathematical resource. Importance is contextual. In Roman society, unlike ours, the capacity to remember numbers was of little importance, because numerical information had little bearing on the cognitive issues of the time.

From an epistemological standpoint, the introduction of the factor of importance (however necessary) poses potential difficulties. For in theory we may well then confront the situation of two distinct and occasionally disagreeing epistemic methods or policies, one of which is more effective for resolving problems in general, and the other of which is more effective for resolving our specifically important problems. In this unwelcome situation, we would be driven to a rather uncomfortable choice by the duality of standards of effectiveness. (This is a situation sometimes actually encountered in natural science, when new methods permit the resolution of newly opened problems without being as effective as the established methods at handling the old problems.) In cognitive contexts, cost-effectiveness assessment is no easier than elsewhere.

IS IMPORTANCE OBJECTIVE?

Is importance something altogether subjective? It is a purely personal issue that lies wholly in the eyes of the beholder? The answer is emphatically negative. Things do not become important merely because people attach importance to them. Matters that bear on the preservation of human life—medicine or nutrition, for example—possess an importance quite apart from any particular individual's view of the matter. Regardless of anyone's personal, idiosyncratic inclinations, they *deserve* people's attention because their very survival is at stake. Again, the importance of calculus for the study of physics does not hinge on people's wishes or beliefs (the yearning of generations of students to the contrary notwithstanding). Things are not made important by people thinking them to be so, any more than something becomes, say, dangerous by people thinking it to be so.[30]

Moreover, it deserves note that the cognitive importance of issues bears no fixed relation to the extent to which people find them to be interesting.[31] Being interesting is always simply a matter of what people happen to be intrigued by; it does indeed lie in the eyes of the beholder. But importance is something that inheres in the nature of things rather than in our thoughts. An extremely interesting subject can be relatively unimportant in the larger scheme of things. (The history of human diversions—games like chess or contract bridge, for example—shows that things can be extremely interesting for us without being very important in themselves.) People find interesting those things that bear closely on their own doings or dealings, or those of people they know about. (Gossip about the scandalous activities of prominent personalities is invariably deemed interesting.) Importance need not enter into it at all.

Cognitive importance must accordingly be recognized as being objective in at least two critical regards. (1) *Unconditional importance* is objective because of its inherently normative nature: it pivots not (just) on what matters for what we do care about, but on what matters for what we should care about—on what our aims ought to be as the sort of beings we are, rather than merely on what our aims actually happen to be. (2) *Conditional importance* is also objective, though now in a hypothetical rather than categorical way. Consider, for example, the conditional thesis, If someone (you or I or anybody) wants to learn physics, then it is important that they should learn calculus. Clearly, this too states a perfectly objective fact. Whether you actually want to learn physics or not is your business. But the importance of advanced mathematics for physics is an altogether impersonal circumstance, a matter of objective fact, much like having a steady hand is important for a surgeon. Conditional importance pivots on the perfectly objective circumstance that something will go amiss as regards some specified function or end if we do not pay due and proper heed to the factor at issue.

Even as it is foolish to think that what you don't know can't hurt you, so it is comparably foolish to think that something people do not deem important is consequently not so. One can err by mistakenly ascribing a false importance to things or by injudiciously according less importance to them than is their due.

THE ROLE OF IMPORTANCE IN ECONOMIC RATIONALITY

In cognitive and practical matters alike, rationality calls above all for the appropriate and sensible allocation of effort. And no aspect of this rational economy of effort is more crucial than giving just due to the things that are important. With virtually any issue or activity we can and should ask; How much time, money, effort, and concern does it deserve? Whether we are planning the activities of a day, a curriculum of studies, the topical coverage of an encyclopedia, or the apportionment of a budget, the issue of relative importance arises. With rational people, importance and resource expenditure stand coordinated in the management of their affairs. Rationality demands specifically that we allocate to matters a share of attention and resources proportionate with their actual importance, expending on an activity no more resources (in terms of time, energy, effort, money, etc.) than its correlative ends are worth.

The conditions of life present us with resource limitations on every side. And with every sort of end-oriented activity, we must divide the pie in a reasonable way: too much for one component means too little for the others. Importance is a key factor for rationality in its concern for returns on investment. A serious imbalance between importance and allocation is not just inappropriate, but actually irrational. (Keeping oneself informed about current events is important, but the person who, in ordinary circumstances, spends 90 percent of his waking hours at this task is being irrational.) In cognitive matters, as elsewhere, even sensible ends can be pursued unintelligently.

Importance accordingly reflects a major economic aspect of rationality. The intelligent allocation of effort, and of resources in general, is a critical factor for rationality at large. Then too, there is the cognate principle that it is not rational to expend limited resources that would yield results of greater value if expended in some other direction. (Importance affords a vivid illustration of how economic rationality is essential to rationality at large.)

FIVE

Induction, Simplicity, and Cognitive Economy

SYNOPSIS

(1) Induction is the methodology for effecting our best estimate of the correct answers to various questions whose resolution transcends the sure reach of the facts in hand. (2) The ideas of economy and simplicity are the guiding principles of inductive reasoning, whose procedure is set by the cardinal precept: Resolve your cognitive problems in the simplest, most economical way compatible with an adequate use of the information at your disposal. (3) The rational basis for our inductive simplicity preference lies in considerations of the economic dimension of practice and procedure, rather than in any factual presuppositions or assumptions about the world's nature. This inherent role of economic considerations in induction is crucial for any rational justification of induction. By its very nature, induction affords us the most cost-effective—the economically optimal—means for accomplishing an essential cognitive task. (4) However, while our commitment to inductive simplicity is indeed a matter of methodological (procedural) convenience, neverthe-

less, our reliance on simplicity is not totally devoid of ontological commitments regarding the world's nature. For our *wisdom-of-hindsight* experience with induction also enters into the feed-back process of its overall justification and, indeed, crucially determines—not, of course, that *we perform inductions*—but rather *how we perform them.*

INDUCTION AS COGNITIVE SYSTEMATIZATION

Dictionaries sometimes define *induction* in such terms as "inference to a general conclusion from particular cases." But such inferences—for example, from spaniels eat meat, schnauzers eat meat, corgis eat meat, to all dogs eat meat—illustrate only one particular sort of inductive reasoning. Nor will it do to add merely those further inferences to a particular conclusion that move from effects to causes from the smoke to the fire, say, or from the bark to the dog. For even this does not go nearly far enough. Inferences from sample to population, from part to whole (from the jaws to the entire alligator), from style to authorship, from clue to culprit, from symptom to disease, and the like, are all also modes of inductive reasoning. The characteristic and crucial thing about inductive reasoning is its overreaching the evidence in hand to move to conclusions lying beyond the informative reach of relatively insufficient data.

Consider a question of the form; Are the Fs also Gs? (Are lions carnivores?) The situation we face in seeking an answer is akin to that of a multiple-choice examination, where one can respond:

1. Yes, all of them are.
2. Never—none of them are.
3. No, some are and some aren't.
4. Unable to decide.

This pretty well exhausts the spectrum of major alternatives for a response. Now when in fact all of the observed Fs (over a fairly wide range) are indeed Gs, our path seems relatively clear. Alternative 4 is not an answer; it is an evasion, a response of last resort, to be given only after all else has failed. Alternative 2 is ex hypothesi ruled out in the circumstances. The real choice lies between 1 and 3. And in the circumstances we naturally prefer the former. The governing consideration here is the matter of plausibility, specifically that of uniformity. For alternative 1 alone extends the data in the most natural and straightforward way, seeing that this response alone smoothly aligns the tenor of our general answer with the available information. It is, accordingly, this economical, uniformity-preserving resolution that affords the inductively appropriate answer in the postulated circumstances.

Induction as we standardly practice it in everyday life and scientific inquiry is a matter of question resolution through the optimal systematization of experience—of answering our questions in terms of the most straightforward overall account that we are able to devise to accommodate the facts or presumptive facts that our observation of the world's phenomena (natural or artificially contrived in experimentation) places at our disposal. Induction is thus a fundamentally regulative and procedural resource for inquiry—one that proceeds by way of implementing the injunction, Maximize the extent to which your cognitive commitments are smoothly systematic overall. In the absence of such a principle, or some functional equivalent of it, the venture of rational inquiry via empirical data will not get under way at all.

Induction is accordingly not so much a process of inference as one of estimation; its conclusions are not so much extracted from data as suggested by them. We clearly want to accomplish our explanatory gap filling in the least risky, the minimally problematic way, as determined by plausibilistic best-fit considerations. Induction *jumps* to its conclusion, instead of literally deriving it from the given premisses by drawing the conclusion from them through some extractive process. Long ago, William Whewell put this key point nicely. "Deduction," he wrote, "descends steadily and methodically, step by step: Induction mounts by a leap which is out of the reach of method (or, at any rate, mechanical routine). She bounds to the top of the stairs at once."[32] Of course, any such ampliative leaps beyond the evidence at hand entail further risks. (In this regard, a scientific revolution is like a market crash, when everything comes tumbling down in one great collapse.) Any inductive process is inherently chancy. Induction is always a matter of guesswork, and its results are always at risk to further or better data. But of course what is involved is responsible, rather than wild, guesswork—rational conjecture rather than fanciful speculations. To be sure, we cannot pass by any sort of strict inference or cognitive calculation from the premisses of an inductive argument to its conclusion because (ex hypothesi) this would be a deductive non sequitur. An inductive conclusion (in the very nature of the case) asserts something well above and beyond the information contained in its premisses.[33] Clearly, the standard and paradigmatic mode of *inference*—of actually deriving a conclusion from the information provided by premisses—is actual deduction,[34] and this paradigm does not fit induction smoothly. As one philosopher has felicitously put it, our inductive "conclusions" are "not *derived* from the observed facts, but *invented* in order to account for them."[35]

The problem of generalizing from particular data is often compared to that of tracing the smoothest curve through a given family of points. As Cournot suggested over a century ago: "En général, une théorie scientifique quelconque . . . peut-être assimilée à la courbe qu'on trace d'après une définition mathématique, en s'imposant la condition de la faire passer par un certain nombre de points donnéés d'avance."[36] Since we seek a curve which passes through the given data points, the question of how we are to estimate the form of this curve is to be resolved on inductive principles in the most systematic (uniform, simple, smoothly continuous) way.[37] The fitting process is to be accomplished by a best-fit model, with the standard parameters of systematicity—uniformity, simplicity, and the rest—playing the determinative role. And this typifies how induction works.

Induction is an instrument for question resolution in the face of imperfect information. It is a resource for use by finite intelligences, a process that yields not the *best possible* answers (in some rarified sense of this term) but the *best available* answers, the best that we can realistically manage to secure in the difficult conditions in which we do and must conduct our epistemic labors. Of necessity, its operation is restricted to what lies within our cognitive reach: it obviously cannot deal with issues that might lie beyond our conceptual horizons (as quantum electrodynamics lay beyond those of the physicists of Newton's day). The available answers at issue have to be found within some limited family of alternative possibilities lying within our intellectual horizons. Induction is not the instrument of a magical alchemy that mysteriously transmutes ignorance into knowledge; it is a mundane and realistic tool for doing the best we can in the epistemic circumstances in which we do—or with reasonable further exertion can—find ourselves.

Induction should accordingly be viewed as a cognitive process, or family of methods, for arriving at our best *estimate* of the correct answers to pressing questions, whose resolution transcends the reach of the facts in hand. Given the information transcendence at issue in such truth estimation, we cannot avoid realizing that an inductive procedure does not guaranteed the truth of its products. Indeed, if the history of science has taught us any one thing, it is that the best estimate of the truth that we can make at any particular stage of the cognitive game is in general subsequently seen, with the wisdom of hindsight, as being far off the mark. Induction opts for simplicity, and its characteristic flaw is thus oversimplification. Nevertheless, the fact remains that the inductively indicated answer does afford our best available estimate of the true answer, in the sense that its adoption represents the best available way of resolving our cognitive problems with the materials at hand.

INDUCTION AND COGNITIVE ECONOMY: THE ECONOMIC RATIONALE OF SIMPLICITY PREFERENCE

Induction is a matter of projecting our cognitive commitments just as far beyond the data as is necessary to get answers for our questions, staying as close to the data as possible while proceeding under the aegis of established principles of inductive systematization: simplicity, harmony, uniformity, and the rest. The ideas of economy and simplicity are the guiding principles of inductive reasoning, whose procedure is that of the precept, Resolve your cognitive problems in the simplest, most economical way compatible with an adequate use of the information at your disposal. Induction proceeds by way of constructing the most economical structures to house the available data comfortably. It seeks to discern the simplest overall pattern of regularity that can adequately accommodate our information regarding cases in hand, and then projects them across the entire spectrum of possibilities in order to answer our general questions. Induction is a process of implementing the general idea of cognitive economy by building up the simplest structure capable of resolving our cognitive problems.

As a fundamentally inductive endeavor, scientific theorizing accordingly involves the search for, or the construction of, the least complex and most straight-

Induction, Simplicity, and Cognitive Economy

forward theory structure capable of adequately accommodating the currently available data. The key principle is that of economy of means for the realization of given cognitive ends, and the ruling injunction is that of cognitive economy—of getting the most effective answer we can with the least effect of complication. Complexities cannot be ruled out, but they must always pay their way in terms of increased systemic adequacy!

It has long been recognized that simplicity must play a prominent part in the methodology of science, constituting is a crucial cognitive value and a paramount factor in inductive reasoning. There is as widespread agreement as there ever is in such foundational matters on the principle that simple hypotheses enjoy a preferred status. But when one presses the question of validating this simplicity preference, one meets with discord and disagreement. The matter becomes far less problematic however, once one approaches it from a methodological rather than a substantive point of view. Henri Poincaré has observed that:

[Even] those who do not believe that natural laws must be simple, are still often obligated to act as if they did believe it. They cannot entirely dispense with this necessity without making all generalization, and therefore all science, impossible. It is clear that any fact can be generalized in an infinite number of ways, and it is a question of choice. The choice can only be guided by considerations of simplicity. . . . To sum up, in most cases every law is held to be simple until the contrary is proved.[38]

These observations are wholly right-minded. As cognitive possibilities proliferate in the course of theory-building inquiry, a principle of choice and selection among alternatives becomes requisite. And here economy and its other systematic congeners—simplicity, uniformity, and the rest—are the natural guidelines. To be sure, whether the direction in which they point us is actually correct is something that remains to be seen. But they clearly afford the most natural and promising starting point. The simplest feasible resolution of our problems is patently that which must be allowed to prevail, at any rate pro tem, until such time as its untenability becomes manifest and complications force themselves upon us. Where a simple solution will accommodate the data at hand, there is no good reason for turning elsewhere. It is a fundamental principle of rational procedure, operative just as much in the cognitive domain as anywhere else, that from among various alternatives that are anything like equally well qualified in other regards, we should adopt the one that is the simplest, the most economical—in whatever modes of simplicity and economy are relevant.

In induction we exploit the information at hand to answer the questions in the most straightforward (economical) way. Suppose, for example, that we are asked to supply the next member of the series 1, 2, 3, 4, . . . We shall straightaway respond with 5, supposing the series to be simply that of the integers. Of course, the actual series might well be 1, 2, 3, 4, 11, 12, 13, 14, 101, 102, 103, 104, ... And then the correct answer will eventuate as 11 rather than 5. Though we cannot rule such possibilities out, they do not deter our inductive

Induction, Simplicity, and Cognitive Economy

proceedings. The inductively appropriate course lies with the production rule that is the simplest answer: Add 1 to the number you've just produced. In induction, we proceed to answer questions by opting for the simplest resolution that meets the conditions of the problem. And we do this not because we know a priori that this simplest resolution will prove to be correct. (We know no such thing!) We adopt this answer, provisionally at least, just exactly because this is the simplest, the most economical way of providing a resolution that does justice to the facts and to the demands of the situation. We recognize that the possibilities exist but ignore them pro tem, exactly because there is no cogent reason for giving them favorable notice *at this stage.*

In inductive situations we are called on to answer questions whose resolution is beyond the reach of information at hand. We simply have to transcend the data. And we do this by projecting our problem resolutions along the lines of least resistance. We try to economize our cognitive effort. We use the simplest workable means to our ends exactly because the others are harder to use. Whenever possible, we analogize the present case to other similar ones, because the introduction of new patterns complicates our cognitive repertoire. We use the simplest viable formulations because they are easier to remember and simpler to use. Insofar as possible, we try to ease the burdens we pose for our memory (for information storage and retrieval) and for our intellect (for information processing and calculation). We favor uniformity, analogy, simplicity, and the like because that lightens the burden of cognitive effort. When other things are anything like equal, simpler theories are bound to be economically more advantageous. We avoid needless complications whenever possible, because this is the course of an economy of effort. And just herein lies the justification of induction. For by its very nature induction affords us the most cost-effective—the economically optimal—means for accomplishing an essential cognitive task. With cognitive as with physical tools, complexities, disuniformities, abnormalities, and so on, are complications that exact a price, departures from the easiest resolution that must be motivated by some appropriate benefit, some situational pressure. Accordingly, the rationale of our inductive praxis is a fundamentally economic one.

Galileo wrote: "When therefore I observe a stone initially at rest falling from a considerable height and gradually acquiring new increases of speed, why should I not believe that such increments come about in the simplest, the most plausible way?"[39] Why not indeed? Subsequent findings may, of course, render this simplest position untenable. But this recognition only reinforces the stance that simplicity is not an inevitable hallmark of truth (*simplex sigillum veri*), but merely a methodological tool of inquiry—a guidepost of procedure. When something simple accomplishes the cognitive tasks in hand as well as some more complex alternative, it is foolish to adopt the latter. After all, we need not presuppose that the world somehow is systematic (simple, uniform, and the like) to validate our penchant for the systematicity of our cognitive commitments. Our striving for cognitive systematicity in its various forms persists even in the face of complex phenomena: the commitment to simplicity in our account of the world remains a methodological desideratum regardless of how complex or untidy the world may turn out to be.

It is the universal practice in scientific theory construction, when other things are anything like equal, to give preference to

- one-dimensional rather than multidimensional modes of description,
- quantitative rather than qualitative characterizations,
- lower- rather than higher-order polynomials,
- linear rather than nonlinear differential equations.

In each case, the former is somehow simpler than the latter alternative. To be sure—efforts to the contrary notwithstanding—no theoretician and no philosopher has managed to provide an adequate *substantive* characterization of simplicity, answering to the formula X is simpler than Y if they stand to one another in a relation of just such-and-such a descriptive sort. But a methodological, or procedural, characterization is something far easier to come by. The comparatively simpler is simply that which is easier to work with, which overall is the more economical to operate when it comes to application and interaction. Simplicity on such a perspective is a concept of the practical order, pivoting simply on being more economical to use—that is, less demanding of resources.

The ideas of economy and simplicity are the guiding principles of inductive reasoning. The procedure is that of the precept, Resolve your cognitive problems in the simplest, most economical way compatible with an adequate use of the information at your disposal. Our penchant for simplicity is easy to justify on grounds of economy. If one claims a phenomenon to depend not just on certain distances and weights and sizes, but also (say) on such further factors as temperature and magnetic forces, then one must design a more complex data-gathering apparatus to take readings over this enlarged range of physical parameters. Or again, in a curve-fitting situation, compare the thesis that the appropriate function is linear with the thesis that it is linear up to a point and sinusoidally wavelike thereafter. Writing a set of instructions for checking whether empirically determined point coordinates fit the specified function is clearly a vastly less complex—and so more economical—process in the linear case.

In inductive reasoning we constantly make use of organizational principles for the structuring of our information: subsumptive classification schemes, connecting laws, coordinating analogies. All of these are means for the assimilation of given cases to general patterns. All such instruments for storing information, recovering it, processing it, and putting it to work, are one and all means for the cost-effective handling of information in the service of resolving our questions. Throughout, operating cost-effectiveness and inductive adequacy run hand in hand. Ease of operation—economy, in brief—is the touchstone of induction, whose guiding idea is just that of employing the normal economic means to adequately serve our cognitive ends.

The impetus of simplicity has one other important ramification. In life we must not only solve problems but also learn. Yet we must learn to walk before we can run, and to solve simple problems before we can solve complicated ones. In simplifying and indeed oversimplifying our problems, we adopt a good strategy for learning. Induction is on the side of cost-effectiveness economics in our cognitive operations.

THE METHODOLOGICAL ASPECT OF INDUCTIVE ECONOMY

On such a view, inductive systematicity is best approached with reference, not to reality as such, or even to our conception of it—but rather, more accurately, to the ways and means we employ in conceptualizing it. Simplicity preference (for example) is based on the strictly method-oriented practical consideration that the simple hypotheses are the most convenient and advantageous for us to put to use in the context of our purposes. There is thus no recourse to a substantive (or descriptively constitutive) postulate of the simplicity of nature; it suffices to have recourse to a regulative (or practical) precept of economy of means. And the pursuit of cognitive systematicity is ontologically neutral. It is noncommittal on matters of substance; it merely reflects the procedure of conducting our question-resolving endeavors with the greatest economy. In inquiry as elsewhere, a principle of least effort predominates; rationality enjoins us to employ the maximally economic means to the attainment of chosen ends. Such an approach constitutes a theoretical defense of inductive systematicity, which in fact rests on practical considerations.

Simpler (more systematic) answers are more easily codified, taught, learned, used, investigated, and so on. In short, they are more economical to operate. In consequence, the regulative principles of convenience and economy in learning and inquiry suffice to provide a rational basis for systematicity preference. Our penchant for simplicity, uniformity, and systematicity in general is not a matter of a substantive theory regarding the nature of the world but one of search strategy—of cognitive methodology. In sum, we opt for simplicity (and systematicity in general) in inquiry because it is teleologically effective for the more cost-efficient realization of the goals of the enterprise, for the parameters of inductive systematicity—simplicity, uniformity, regularity, normality, coherence, and the rest—all represent practical principles of cognitive economy.[40] They are labor-saving devices for the avoidance of complications in the course of our endeavors to realize the objects of inquiry. The rationale of simplicity preference is straightforward. It lies in the single word *economy*. The simplest workable solution is that which is the easiest, most straightforward, most inexpensive one to work with. It is the very quintessence of foolishness to expend greater resources than are necessary for the achievement of our governing objectives. And by its very nature, induction affords us the most cost-effective—the economically optimal—means for accomplishing an essential cognitive task.

In all contexts, cognitive ones included, the rational agent opts for the simplest workable solution. We certainly do not do this because we know a priori that the simple answers are bound to prove correct. Rather, we adopt the simplest viable solutions until further notice (i.e., until they may prove to be no longer viable) just exactly because they are the simplest—just because there is by hypothesis no good reason whatsoever for resorting to a more complex possibility. The rational basis for our inductive simplicity preference lies in considerations of the economic dimension of practice and procedure, rather than in any factual supposition about the world's nature.

It is indeed economy and convenience that determine our regulative predilection for simplicity and systematicity in general. Our prime motivation is to get by with a minimum of complication, to adopt strategies of question resolution that enable us, among other things: (1) to continue with existing solutions unless and until the epistemic circumstances compel us to introduce changes (uniformity), (2) to make the same processes do as great a variety of scientific tasks as possible (generality), and (3) to keep to the simplest process that will do the job (simplicity). Such a perspective combines the commonsensical precept, Try the simplest thing first, with this principle of burden of proof: Maintain your cognitive commitments until there is good reason to abandon them.[41] It clearly makes eminent sense to move onward from the simplest (least complex) available solution to introduce further complexities when and as—but only when and as—they are forced upon us.

From this perspective, then, simplicity preference emerges as a matter of simplification of labor, a matter of the intellectual economy of cognitive procedure. Why use a more complex solution where a simple one will do as well? Why depart from uniformity? Why use a new, different solution where an existing one will serve? The good workman selects his tools with a view to (1) their *versatility* (power, efficacy, adaptability, and the like), and (2) their *convenience* (ease of use), and other similar factors of functional adequacy to the task in hand.[42] Simplicity preference, accordingly, emerges as a means of implementing the precepts of economy of operation in the intellectual sphere. Its initial advantages are not substantive/ontological but methodological/pragmatic in orientation. The crucial fact is that simplicity preference is a cognitive policy recommended by considerations of cost effectiveness; in the setting of the cognitive purposes at issue, it affords a maximally advantageous inquiry mechanism.

The informative benefits of knowledge relate to the intrinsic value of having or using information. Besides these costs and benefits of knowledge, there are also certain structural costs and benefits implicit in information—economic features of the body of (purported) knowledge itself, in contrast to the informative services that it provides us. These structural values of information embrace factors like generality, comprehensiveness, completeness, uniformity, simplicity, coherence, and elegance—or on the negative side, fragmentedness, limitedness, incompleteness, disuniformity (eccentricity), complexity, and dissonance. Benefits here relate to the intrinsic economy or elegance, and costs to the intrinsic complexity and cumbersomeness of the information at issue (see table 5).

The penchant for inductive systematicity reflected in the structural dimension of information is simply a matter of striving for economy in the conduct of inquiry. It is governed by an analogue of Occam's razor—a principle of parsimony to the effect that needless complexity is to be avoided. Given that the inductive method, viewed in its practical and methodological aspect, aims at the most efficient and effective means of question resolution, it is only natural that our inductive precepts should direct us toward the most systematic, and thereby economical, device that can actually do the job at hand. Our systematizing procedures pivot on this injunction always to adopt the most economical (simple, general, straightforward, etc.) solution that meets the demands of the situation. The root principle

TABLE 5
Structural Advantages and Disadvantages of Information

Positivities	Negativities
Comprehensiveness (informativeness)	Narrowness (uninformativeness)
Completeness	Incompleteness
Probability	Improbability
Generality	Particularity
Coherence (fit, inner consistency)	Incoherence (dissonance)
Simplicity	Complexity
Uniformity	Disuniformity
Connectedness	Disjointedness
Elegance	Cumbersomeness
Robustness	Fragility

of inductive systematization is the axiom of cognitive economy: *complicationes non multiplicandae sunt praeter necessitatem.* The other-things-equal preferability of simpler solutions over more complex ones is thus obvious enough: they are less cumbersome to store, easier to take hold of, and less difficult to work with.

THE ONTOLOGICAL RAMIFICATIONS OF SIMPLICITY

But the aim of inquiry is to get at the truth of things, and is there any reason to think that simpler theories have a better prospect of being true? Clearly there are difficulties here. Does nature exhibit a penchant for simplicity? Surely not. We cannot say, solely on the basis of general principles of some sort, that this world—the real world as such—must of necessity be a simple one. Nor is there any real need for doing so.

Hans Reichenbach has written: "Actually in cases of inductive simplicity it is not economy which determines our choice.... We make the assumption that the simplest theory furnishes the best predictions. This assumption cannot be justified by convenience: it has a truth character and demands a justification within the theory of probability and induction."[43] This perspective is gravely misleading. What sort of consideration would possibly justify the supposition that "the simplest theory furnishes the best predictions"? Any such belief is surely inappropriate. Induction with respect to the history of science itself—a constant series of errors of oversimplification—would soon undermine our confidence that nature operates in the way we would deem the simpler. On the contrary, the history of science is a highly repetitive story of simple theories giving way to more complicated and sophisticated ones. The Greeks had four elements; in the nineteenth century, Mendeleev had some eighty; we nowadays have a vast series of stability states. Aristotle's cosmos had only spheres; Ptolemy's added epicycles; ours has a virtually endless proliferation of complex orbits that only supercomputers can approximate. Greek science could be transmitted on a shelf of books; that of the Newtonian age required a roomful; ours requires vast storage structures filled not only with books and journals but with photographs, tapes, floppy disks, and so on. Of the quantities nowadays recognized as the fundamental constants of physics, only one was contemplated in Newton's physics, the universal gravitational constant. A second was added in the nineteenth century, Avogadro's constant. The remaining six are all creatures of twentieth century physics: the speed of light (the velocity of electromagnetic radiation in free space), the elementary charge, the rest mass of the electron, the rest mass of the proton, Planck's constant, and Boltzmann's constant.[44] It would be naive—and quite wrong—to think that the course of scientific progress is one of increasing simplicity.

A methodologically based approach to the rationalization of inductive systematicity on grounds of economy accordingly is not predicated on presupposing any sort of ontological linkage between simplicity and (probable) truth. In the first instance, at any rate, the thesis that nature is a system is seen as a merely regulative principle, a guideline of procedure. The basis of simplicity preference is thus methodological or epistemological rather than substantive or ontological.[45] The pivotal role that science assigns to cognitive virtues like simplicity, coherence, systematicity, fecundity, symmetry, and generality on first view appears as problematic and question begging. It is only when we turn to the methodological standpoint of procedural economy that everything falls easily and naturally into place.

It is important, however, to distinguish economy of means from economy of product—methodological from material economy. Simple tools or methods can, suitably used, create complicated results. A simple cognitive method, such as trial and error, can ultimately yield complex answers to difficult questions. Conversely, simple results are sometimes brought about in complicated ways. A complicated method of inquiry or problem solving might yield easy and uncomplicated problem solutions. Our commitment to simplicity in scientific inquiry does not, in the end, prevent us from discovering whatever complexities are actually there.

To be sure, this methodological/procedural tale is not the whole story. There is also, in the final analysis, a substantive aspect to the matter of induction's justification. Our intellectual tastes (for simplicity, elegance, etc., as we naturally construe these ideas) are, like our physical tastes (palatability), the product of evolutionary pressure to prioritize those things that work—that prove effective and are thus survival conducive. The evolutionary aspect of our cognitive mechanisms assures the serviceability of the cognitive values we standardly invoke as effective conditions of adequacy for the substantiation of information. The presence of positive values in information enhances the utility of that information, not because nature is benign or because a preestablished harmony is at work, but because evolution—both biological and cultural—so operates as to assure alignment. Nature exacts its penalties for ineffectiveness, and evolutionary pressure toward cost effectiveness assures an inherent connection between functional adequacy and temporal survival. The evolutionary realities assure an important role for economic considerations in the theory of knowledge. There is a tight linkage between a cognitively advantageous economy of intellectual effort and a biologically advantageous economy of physical effort.

Take memory, for example. In general, the information we most urgently need is just the information that we most frequently use. And repeated use renders if familiar, impressing it more deeply into our memory with each repetition. The transit from need to use to familiarization to availability is assured by a convenient chain of statistical linkages that reflects the modus operandi of our memory. Evolution has so arranged matters that the recentness of our experience is significant for short-term memory and its frequency for long-term memory. As this example indicates, cost effectiveness is inevitably coordinated with the implicit rationality of evolutionary process by virtue of the survival conduciveness of arrangements that represent efficient ways of using limited resources.

Moreover, the development of our cognitive methods through rational selection also plays a key role in this connection. The process of cognitive evolution so unfolds as to assure the coordination of convenience with effectiveness, for a process of rational selection is at work to support the retention, promulgation, and transmission of those cognitive resources that prove themselves effective in operation. The burden of this evolutionary argument is not biological survival. The point is not that in a niche rivalry between Austere Simplifiers and Byzantine Complicators the former will eventually *displace the latter biologically*, but rather that, if both are intelligently persistent, the former will eventually *outdistance the latter epistemologically*, simply through the circumstance that they perform in more cost-efficient ways. Cultural dominance in a community of intelligent agents, not biological selection alone, plays a crucial part in the development of our cognitive instrumentalities.

Accordingly, our reliance on induction is not wholly methodological but also has an ontological, realistic aspect, in that we learn by experience how to practice induction—that is, how to go about the process of a conservation of effort. Trial and error—that is, the course of experience—constrains us to bring methodological/procedural economy into alignment with substantive/ontological economy in our cognitive operations. In particular, the reification of the mechanisms of our simplest explanations (unobservable entities and the like) affords a powerful heuristic. It is the (empirically confined) efficacy of such a process that provides the ultimate justification of such a realistic approach. We are well advised to accept unobservable entities not because their existence is somehow confirmable in observation (which ex hypothesi it is not) but because experience shows that a methodology of inquiry predicated on such a simplifying assumption in the end affords our most efficient and effective resource.

The justification of relying on simplicity in the pursuit of our cognitive affairs will initially rest on an essentially instrumental basis. We incline to initially prefer the optimally systematic (simple, uniform) alternative, because this is the most economical, the most convenient, thing to do. But we persist in this course because experience shows the utilization of such economical methods to be efficient, to be optimally cost effective (relative to available alternatives) for the realization of the task. The regulative principles and procedures at issue in our inductive practices are ones whose legitimation lies in their being pragmatically retrovalidated through a demonstrated capacity to guide inquiry into successful channels.

Induction thus emerges as a matter of the pursuit of systemic economy in the cognitive sphere. Inductive simplicity and systematicity inhere in a regulative ideal of inquiry correlative with the procedural injunction, So organize your knowledge as to impart to it as much systematic structure as you possibly can! A cognitive venture based on the quest for simplicity and systematicity, while at first merely hopeful, is ultimately retrovalidated in experience by the fact that its pursuit enables us to realize the fundamental aims and purposes of the cognitive enterprise more efficiently than the available alternatives. Initially, the pivotal issue is simply the matter of our convenience in doing what must be done to serve our purposes. The whole ontological question of the systematicity of nature can safely be left to await the results actual use of inductive processes. No prior presuppositions are needed in this regard.

Such a process of what might be characterized as *retrojustification* underwrites, ex post facto, the substantive conclusion that a methodology of inquiry geared to systematicity preference is efficient—that it accomplishes the ends at issue in the cognitive enterprise with due economy of means. For the pivotal fact is not that (as Reichenbach puts it) "we make the assumption that the simplest theory furnishes the best predictions"—an assumption obviously ill-advised in the light of experience—but that plausible expectation preindicates and actual experience retrojustifies the supposition that a process of inquiry that proceeds on this basis is comparatively efficient in the realization of our cognitive goals. The point is not that the simplest alternative demonstrably affords (or is more likely to afford) successful predictions of and interventions in nature, but that a policy of inquiry that embodies simplicity preference emerges as a relatively effective epistemological policy. In the end, our inductive preference for simplicity and systematicity finds its final justification in the fact that it affords an effective search policy for serviceable truth estimates regarding answers to our questions; it is a productive strategy of inquiry rather than an immediate index of truth.

The crux is that we ultimately learn by experience (and thus through inductive reasoning itself) how to accomplish our inductive business more effectively. Our recourse to induction—that is, *that* we proceed by its means—is justified instrumentally. For induction is a self-improving process. Experience can itself teach us what ways of interpreting the fundamental procedural ideas of inductive practice (simplicity, conformity, generality, and the rest) can lead to improved performance in the transaction of our inductive business. By a cyclic feedback process of variation and trial, we learn to do induction more effectively. *Economy* and *convenience* play the crucial pioneering role in initially justifying our practice of inductive systematization on procedural and methodological grounds. But, in their turn, the issues of *effectiveness* and *success* come to predominate at the subsequent stage of retrospective revalidation ex post facto. And the question of the seemingly preestablished harmony coordinating of these two theoretically disparate factors of convenience and effectiveness is ultimately resolved on the basis of evolutionary considerations in the order of rational selection.[46]

And so, while our commitment to inductive simplicity should be seen as a matter of methodological convenience within the overall economy of rational procedure, nevertheless, our reliance on simplicity is in the event not totally devoid of ontological commitments regarding the world's nature, for our wisdom-of-hindsight experience with induction also enters into its overall justification and, indeed, crucially determines not *that*, but *how* we perform inductions.

SIX

Economics and the Methodology of Inquiry

SYNOPSIS

(1) Various interesting cases illustrate how economic considerations can clarify otherwise puzzling features of our cognitive affairs. (2) In particular, they explain the phenomenon (illustrated by C. G. Hempel's raven paradox) that logically equivalent statements can function very differently in scientific contexts. (3) Nelson Goodman's renovated riddle of induction also loses much of its puzzlement when one adopts an economic point of view. (4) The Popperian contentions about the role played by considerations of generality in scientific contexts can also be clarified from an economic perspective. (5) and (6) Novelty tropism and the role of symmetry in scientific reasoning can also be given an economic explanation. (7) J. M. Keynes' problem of whether to decide issues now, subject to imperfect evidence, or to postpone deciding and secure more information is also an issue that cries out for an economic approach. (8) As such deliberations indicate, the methods and procedural principles of scientific

practice are often best understood and explained on an economic basis.

INTRODUCTION

This chapter will offer some illustrations of how the theory of inquiry can be helpfully illuminated from an economic point of view, presenting some case studies to show how epistemological issues can be clarified on the basis of economic considerations.

HEMPEL'S PARADOX OF THE RAVENS

A substantial literature has grown up over a puzzle of inductive reasoning first posed by Carl G. Hempel in 1946, the so-called paradox of the ravens.[47] It is rooted in the observation of deductive logic that law statements of the form All X are Y are logically equivalent (by contraposition) to All non-Y are non-X, so that, for example, the statement All ravens (R) are black (B) is logically interdeducible with All non-black objects (non-B) are non-ravens (non-R). But a confirming instance of the former law, namely a black raven, is something very different from a confirming instance of the latter, say a white tennis shoe, which no one would consider instancing in support of All ravens are black. Given the *deductive* equivalence of these statements, why should it be that in *inductive* contexts one is prepared to accept black ravens as confirming instances of the original law claim, but not white tennis shoes? How can confirming evidence bear differently on logically equivalent generalizations?

The situation is rendered graphic by the Venn diagram of figure 3. The emptiness of compartment (2) is quite as equivalently determined by going to the Rs = [(2) + (3)] and seeing that only (3)s are encountered as by going to the non-Bs = [(1) + (2)] and seeing that only (1)s are encountered. Either way, we simply determining the emptiness of compartment (2). There seems to be no cogent theoretical reason for granting one of these confirming instances a preferred status over the other. Why, then, do these two approaches seem to differ so drastically from an inductive standpoint?

The answer is not far to seek. The fact is that there is a crucial economic difference between the two approaches. To check the emptiness of (2) by the natural

FIGURE 3
A Venn Diagram for the Raven Paradox

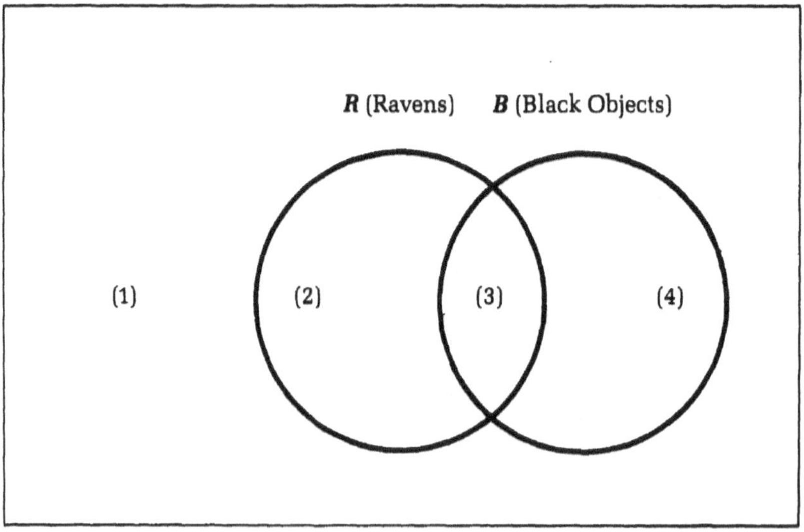

approach (via Rs) means going to the Rs = [(2) + (3)] and checking their color. To establish it by the unorthodox alternative approach (via non-Bs) means going to the non-Bs = [(1) + (2)] and checking their type. A few assumptions may be made to delineate the quantitative structure of the situation:

1. The number of Rs is approximately 10^8.
2. The number of non-Bs is approximately 10^{40}.
3. The average cost of finding an R is around $1.
4. The average cost of finding a non-B is around .1 cent ($.001).
5. The cost of determining the blackness of a given R is approximately 1 cent ($.01).
6. The cost of determining the ravenhood of a given non B-in-hand is approximately .1 cent ($.001).

Nothing much hinges on the particular numbers here. Specifically regarding 2, there is little point in contemplating some Eddington-reminiscent estimate of the number of molecules in the universe; all that we need for present purposes is that the number of identifiable objects in the world be rather big.

Let it be assumed further that to get adequate statistical control of a population of size X, we need to have a sample of roughly the size $\sqrt{X}$. (Again, the result will prove highly insensitive to the exact character of this control assumption, so that its details are not worth quarrelling over.) Two courses of action now lie before us: The first approach is to find a suitable number of ravens and check their blackness:

Cost: $10^4 \times \$1.00 = \$10{,}000$.

The second approach calls for finding a suitable number of nonblack objects and checking their ravenhood:

Cost: $10^{20} \times \$.001 = \$1 \times 10^{17}$.

The difference in cost is enormous. Once we take the actual *process* of verification into account, and consider its inevitable economic aspect, there is a striking economic difference between these two cases. And this difference puts black ravens and white tennis shoes on an altogether different plane. To be sure, both black ravens and white tennis shoes in some degree confirm the generalization at issue. But they do so to a very different extent and make contributions of very different value. If someone hands us a confirmed black raven, he has contributed one hundredth of 1 percent to the cost of the overall project of recurring adequate verification. But if he hands us a confirmed white tennis shoe, his contribution is vanishingly small owing to the comparative vastness of the population at issue (nonblack nonravens). The natural course of approach is thus massively cheaper; with it, an outlay of given size goes enormously further toward adequacy, yielding vastly larger returns by way of confirmation or disconfirmation. Reliance on the natural instance wins hands down in cost effectiveness as an inductive strategy. Heed of the economic dimension accordingly renders good service here.

Again, suppose that we are investigating the question of whether a certain object x has the property F in circumstances where we cannot directly resolve the issue, Is x an F? That is, we must proceed by determining x's status in regard to some other properties, say G or H, which we know to be relevant to F. And suppose further that we know on the basis of background considerations that things are distributed as follows with respect the properties at issue:

	F	non-F		F	non-F
G	10%	90%	H	55%	45%
non-G	90%	10%	non-H	45%	55%

Then if we learn x's G-status, we know something quite useful: if it lacks G, then it is very likely to be an F, and if it doesn't, it isn't. But if we learn x's H-status, then we are not very far ahead, since either way, H-possessing things are just about as likely to be F as not. We are, in the circumstances, much better advised to worry about x's status vis-à-vis G rather than H.

As the game of Twenty Questions makes clear, there are efficient and inefficient ways of constructing the question sequence that constitutes an investigation. Since the resolution of questions through rational inquiry is a resource-consuming process, the issue of cost effectiveness is clearly going to arise in constituting the question sequence (course of investigation) through which a research program proceeds. In general, it is not through abstract general principles but by a sensible heed of economic considerations that an effective program of inquiry can be devised.

GOODMAN'S GRUE PARADOX

Let us also reexamine Nelson Goodman's well-known revised riddle of induction from an economic perspective. In an influential essay of 1953, Nelson Goodman propounded a puzzle about inductive inference, which occasioned an enormous literature over the ensuing years. Goodman's puzzle emerged from his definition of two somewhat unorthodox color concepts:

> Grue = a color that is examined before the temporal reference point t and is green or is not examined before t and is blue. (Here, the temporal reference point t is an otherwise arbitrary moment of time that is not in the past.)
>
> Bleen = a color that is examined before the temporal reference point t and is blue or is not examined before t and is green.

If we do our inductive reasoning on the basis of this color taxonomy, we shall obtain excellent inductive support for the thesis that all emeralds will eventually (after t) have the appearance we have standardly indicated by the description blue, since all emeralds that have been examined so far have been found to be given in color. Given this basis, our normal inductive expectations regarding the future would be totally baffled. We arrive at a Hume-reminiscent scepticism about induction from a totally un-Humean point of departure, for we continue to reason inductively—albeit with our nonstandard predicates—and take the stance that the future will resemble the past *with reference to them,* so that all our inductively based expectations are turned topsy-turvy, and our world picture turned into chaos.

Various considerations mark this puzzle as a problem that cuts deeper than it might at first appear.

For one thing, no amount of empirical (observational) evidence can help us choose between the abnormal grue/bleen color taxonomy and the normal and accustomed green/blue one. Empirical evidence must, in the nature of things, relate to the past or present, and there is (ex hypothesi) no difference here from the standard color situation, since this time span at issue is, by hypothesis, antecedent to t.

Moreover, one cannot reasonably object that grue and bleen make explicit reference to time (to t), because this only seems so from our own vantage point. From the perspective of the grue-bleeners who actually use this alternate color terminology, it is *our* color taxonomy that appears to be time dependent:

Green = examined before t and grue in appearance or not examined before t and bleen.

Blue = examined before t and bleen in appearance or not examined before t and grue.

The situation is entirely symmetric from their perspective, and what's sauce for the goose is also sauce for the gander as far as any theoretical objections on the basis of general principles go. In general, then, the situation looks to be one of total parity between our color talk and that of the grue-bleeners, with no theoretical advantage available to decide the choice between them and us.

Goodman himself ultimately abandoned the search for a preferential rationale of choice on the basis of theoretical general principles. Instead, he fell back on an appeal to "entrenchment"—the fact that prevailing custom and habit favor our established color language habits. Few commentators have found this resolution convincing.[48]

However, the whole issue wears a very different aspect when approached from the direction of economic considerations. To operate with grue/bleen as one's operative color taxonomy, it would eventually (i.e., after t) not suffice to recall the phenomenological circumstance that the thing looked just like *this* or like *that*, ignoring whether we saw it before or after t. It would not serve to have faithful color photographs of the thing; if we had no record of when they were taken, we could not say what its color was. In the long run (i.e. after t), mere appearance is not enough; timing becomes crucial. The issue of color characterization ceases to be a purely phenomenological one.

With the orthodox green/blue color taxonomy, however, all that ever matters—not only with current perception but also with pictorial records, memory, and even precognition—is the strictly phenomenal issue of the perceived appearance of things. Thus mere experiential confrontation will in itself suffice to teach color concepts. In the standard case, unlike that of grue/bleen it happens that ostension is an adequate apparatus for learning and teaching about colors, because phenomenal appearance is the only issue.[49] Thus green/blue can be handled by purely ostensive surrogates (looking, in point of color, just like this leaf or like that bit of sky), while with grue/bleen there is always the further complication of the time of observation relative to the great divide at t. With regard to matters of appearance, the orthodox case involves simply the ostensively manipulable phenomenology of observation; the unorthodox case also calls for a further chronometric recourse to temporal data. It is, accordingly, a matter of economy (not of chance!) that the "normal" taxonomy is normal, for its use is more straightforward, simpler, and more economical. With standard colors, mere appearances suffice; with Goodman's nonstandard ones, they do not.

It must be emphasized that such a solution does not rest on an invidious comparison of the color taxonomies from the unfairly contrived vantage point of a rival alternative. It is a strictly internal issue of what it takes to operate the conceptual machinery involved in actually applying these color characterizations over time. The operative contrast between these color taxonomies is not drawn from the standpoint of a precommitment to one but simply involves the question of the resources or mechanisms needed for implementing each scheme.

We are now not left (as Goodman was) with the brute fact of entrenchment itself—the circumstance that the orthodox taxonomy is established, accepted, and familiar in contrast to its alternatives. Rather, we are able to deploy fundamental principles to get some insight into *why* it should be the one that has become entrenched; namely, because it is less complex and more convenient (economical!) to operate. Entrenchment is thus rendered operative not as a matter of mere historical happenstance but as a factor that enjoys a perfectly sound rationale on this basis of economy.[50]

It thus emerges that the problem of validating our inductive recourse to the orthodox color taxonomy against its Goodmanian rivals can be resolved on matters of principle rather than such wholly contingent matters of mere fact as Hume's "custom" or Goodman's "entrenchment." To be sure, the currently operative principles are practical—and, in particular, economic—rather than theoretical in their orientation and import. The considerations needed to cut through the Gordian knot are precisely those involved in the economic factors of efficiency and convenience.

GENERALITY PREFERENCE AND FALSIFICATIONISM

Let us examine from an economic point of view yet another important issue in the methodology of inductive inquiry: the significance of *generality*. A convenient starting point is provided by Karl Popper's widely influential insistence that, "It is this interest in the testability of hypotheses which leads . . . to my demand that . . . statements of a high level of universality should be chosen for scrutiny and testing."[51] As Popper sees it, falsifiability is the crux. Scientific rationality requires giving priority in one's investigations to the most general—the most daring and therefore vulnerable—hypotheses among the available possibilities.

Does such an approach make sense? It is certainly true that, from the angle of Popperian falsificationism, the more general theories—those with fewer qualifying complications and qualifications—are more susceptible to falsification and thus (other things equal) are less probable. But does this vulnerability constitute a decisive factor for the methodology of scientific inquiry?

Suppose, for the sake of an example, a scientific problem situation in which we have 110 alternative, hypothetically available, possible solutions, falling into two groups as shown in table 6. Suppose further that 120 units of resources are at our disposal to investigate the matter. We are to put to the test x hypotheses of group A and y hypotheses of group B. How are we to decide on specific values for x and y? If we are true Popperians, we turn directly to group B, where the more vulnerable possibilities are obviously found. But is this reasonable?

TABLE 6
An Assumed Situation of Two Groups of Problem-Resolving Hypotheses Differing in Probability and Generality

Situation	Group A	Group B
Number	10	100
Probability of each hypothesis (on the basis of a best estimate relative to the evidence at hand)	.05	.005
Generality of each hypothesis[a]	20	80
Unit resource cost of testing one of the hypotheses	10	1

[a] For simplicity, we suppose that generality may be ranked on a scale from 0 to 100 (say for the percentage of the theoretically relevant species of individuals to which hypotheses of the range at issue might conceivably apply).

Note that the expected value we will obtain in point of generality is given by the quantity:

$$.05x(20) + .005y(80) = x + .4y.$$

And so, it may plausibly be supposed, the sensible thing to do is to maximize this quantity, subject to the given basic constraints $x < 10$; $y < 100$; and $10x + y < 120$. The result is the neat little linear programming problem depicted in figure 4. Accordingly, $x + .4y$ is maximized with a value of 42 at $(2, 100)$. That is, we are to work through all 100 of the group B hypotheses and also two other (randomly chosen?) group A hypotheses. The high-generality hypotheses definitely wind up in the preferred position here.

FIGURE 4
A Diagrammatic View of the Example

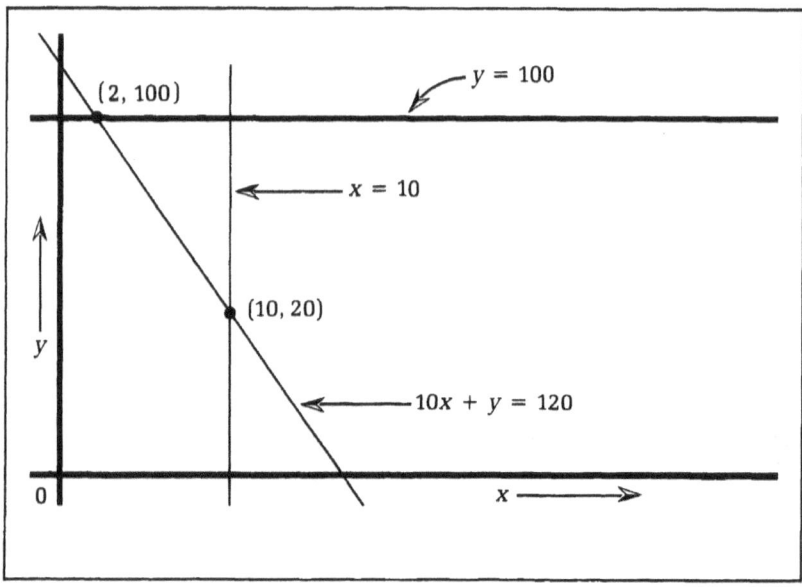

But, of course, had the case been somewhat different, the result would differ too. The fact is that an economically oriented approach is clearly sensible, though wholly undogmatic regarding generality preference. It replaces Popper's purely logical concern for generality for its own sake with an economic/methodological concern for universality relative to cost. If we take this economic approach in basing our decision on the perfectly reasonable precept, Maximize generality subject to the constraints of affordability, then our basic concern is one of cost-benefit assessment—of seeking to optimize returns subject to resource outlays.[52] From this (surely sensible) perspective, it becomes altogether secondary whether priority is given to the more or the less general alternatives. Once due heed is paid to the economic aspects of the matter, the unswerving ideological attachment on the basis of general principles for highly general (or highly specific) hypotheses becomes a luxury we can no longer afford.

To be sure, it is clear that generality—simple extent of reach and range—is going to play a central role in any theory of inquiry where cost effectiveness is a consideration. On the orthodox approach, science is interested in scrutinizing general theses because they are, other things equal, maximally informative. On a Popperian approach, science is generality oriented because general theses are, other things equal, more vulnerable, more falsifiable—more readily testable. On an economically geared approach, the matter stands very differently. Science is generality oriented because, other things equal, general theses are the most cost effective.

Consider the following sequence:

1. All (10) lions in the zoo have tails.
2. All (1,000) lions in the United States have tails.
3. All (100,000) living lions have tails.
4. All (100,000,000) lions who have ever lived have tails.

Let us compare the relative advantages of these generalizations as foci of research. Two considerations will now enter in. First, there is the (previously supposed) fact that to have reasonably good statistical control over a population of size N we must check a sample of (say) $\sqrt{N}$ of that population (given reasonable randomness of selection). second, there is the circumstance that, in processing X individuals of a certain sort, the average cost of the operation decreases with the size of X by a mass-production effect (say with an average unit cost of $1/\log X$, subject to the well-established principle of the economics of production that the cost of the nth unit is proportional to $1/n$). Thus in comparing the four-item sequence above, it becomes clear that we will get a ten-million fold increase in the range of application for a 500-fold increase in expenditure. As this example illustrates, generality is obviously very advantageous in point of cost effectiveness. On the whole, there is much to be said for the Popperian approach—not, however, simply because of its commitment to generality, but because of its economic advantages.

These deliberations indicate that, while from the orthodox approach general hypotheses are the most *informative* (and on a Popperian approach they are the most *testable*), the case is different when we take an economic approach. They are now superior because they combine these desiderata, offering the most information content per unit of effort invested in testing. Generality preference is now not a matter of dogmatic predeliction but something that has a thoroughly intelligible rationale on the basis of economic considerations.

Note, however, that what is at issue in the present deliberations is not the economic advantageousness of selecting a certain hypothesis for *adoption* (or rational acceptance), but rather the quintessentially abductive issue (in C. S. Peirce's terminology) of selecting a certain hypothesis for serious consideration—in particular, for *testing*. The distinction between research-worthy hypotheses and acceptance-worthy credible theories is crucial here. Evidentiation is one thing, investigation worthiness another. Within the orbit of these Popperian deliberations, our focus of concern has to be reoriented to lie with the design of a research program and not with the subsequent issue of theory acceptance.

NOVELTY TROPISM

An economic perspective can also help to account for another otherwise puzzling aspect of scientific methodology: the predisposition of research scientists toward novel and untried theories. In science, a bandwagon effect generally operates to ensure that, once a new theory begins to gain acceptance, it replaces its old rival with surprising rapidity and manages to gain approval to an extent that outstrips its evidential merits. From the very nature of the situation, this penchant toward novelty cannot be explained on evidential grounds. How then can one explain it?

Potentially greater fertility is the pivotal factor here. The new theory may not as yet have strong track-record claims, but it has more promise. The old theory has had a run for its money. What it can do for us by way of providing explanations, suggesting experiments, and the like, has by now been seen. Thus work on the basis of the new theory is more cost effective. Moreover, precisely because the new is more promising, the reward system of science operates in its favor, for it proceeds on the basis that it counts as a bigger contribution to undermine something accepted and well entrenched rather than to squeeze yet another drop of service from a theory that is old hat.

Here too, then, we have another instance where a seemingly questionable aspect of the design of scientific research programs can be accounted for straightforwardly on an economic basis of cost-effectiveness considerations.

MAKING VERSUS POSTPONING DECISIONS

Epistemologists have long been concerned—and frustrated by the question: If an issue has to be resolved in the face of imperfect information, ought we to decide it now as best we can with the evidence in hand, or ought we to postpone a decision and secure more information?[53] It is easy to understand why theorists who like to proceed on the basis of general principles have been reluctant to address this question, for from a purely theoretical point of view, there is no good reason why we should ever have to come to a decision at all as long as our evidence is in any respect incomplete—as it virtually always is. If the penality of delay is zero and is conjoined with a similar costlessness for information enhancement, then nothing is lost by delaying a decision and awaiting developments. Only by taking the practicalistic turn away from pure theory and bringing costs into it can we arrive at a realistic view.

Consider an illustration. Let us suppose that the parameters of the situation are as indicated in table 7, where the benefits of deciding correctly and the penalties of deciding incorrectly are assumed (somewhat unrealistically) to be constant over time. Note that if we deploy the familiar expected-value comparison as our guide, in the manner standard in decision theory, then

TABLE 7
Making Versus Postponing Decisions

Situation	Cost (additional)	Probability of Deciding Correctly	Benefit of a Correct Decision	Loss from a Wrong Decision
Decide now.	0	p	$+B$	$-L$
Decide later (after more inquiry).	C	$p + d$	$+B$	$-L$

EV(now) = $p(B) + (1 - p)(-L)$.
EV(later) = $(p + d)(B) + (1 - p - d)(-L) - C$.
EV(later) > EV(now) iff $d(B + L) > C$.

1. If $C = 0$ (that is, if additional information is cost free), then it clearly *always* pays to postpone the decision.
2. If there is no penalty from a wrong decision (that is, if $-L = B$), then it clearly *never* pays to postpone the decision.
3. If we measure the costs in total-stake units, so that $C = x(B + L)$, then the rule is simply to postpone if $d > x$, that is, if the incremental probability of a correct decision exceeds the proportional increase in cost.

As these relationships indicate, the second-level decision of whether to decide an issue now or to postpone deciding for the sake of further inquiries can be regarded as a straightforward matter of balancing the expectation of costs and benefits involved—in exactly the way in which one standardly looks at such issues in decision theory. Whenever the information needed for an informed assessment of expectations is available, then a plausible decision based on cost-benefit economics is immediately feasible. And this illustrates the general situation. The issue of now versus later is not one of abstract theory but of economic detail. It is worthwhile delaying decisions in order to make them in the light of fuller information as long as—but only as long as—the balance of economic advantage is such as to favor this course.

SYMMETRY ARGUMENTS

Anaximander of Miletus, the early Greek nature philosopher who lived in sixth century B.C., taught that "the earth swings free, held in its place by nothing. It stays where it is because of its equal distances from everything."⁵⁴ Anaximander thus saw the earth as positioned exactly in the middle of an essentially symmetric cosmos. Given this symmetry of its placement, any position-modifying inclination to move toward one extreme would be matched—and thus cancelled out—by an equal and opposite countervailing inclination to move toward its opposite, so that the earth would remain fixed in the middle. The world's symmetry issues in a physical balancing out that leaves the earth freely suspended in the middle of the cosmos.

This appeal to symmetry considerations has ever since been widely accepted as a principle of reasoning in physical science. What is at issue in this sort of reasoning is that things that are just alike in certain respects must behave the same way in these respects, as per the principle, If whatever holds for X in a certain regard R also holds for Y, then X and Y must behave exactly alike in all respects relevant to R. The governing idea is that any difference in behavior must be causally rooted in differences in condition.⁵⁵ In situations of symmetry, equal and opposite tendencies cancel out.

The problem of validating this sort of symmetry/uniformity thesis as a physical principle is clearly a formidable one. How can one possibly be confident that mere chance cannot break symmetry by severing what physical uniformities link together? And this difficulty of expelling chance is nowadays more acute than ever, for once one follows the path of modern physics and accepts genuinely stochastic phenomena—such as the radioactive disintegration of unstably heavy atoms—then one has to come to terms with the fact that the old-line metaphysicians' principle of causality has fallen victim to the physics of chance.

But the matter takes on a very different aspect once we abandon the substantive standpoint of physical operations and regard the matter from an epistemological point of view. Here the issue is not one of legislating for nature (If nature is going to act differently regarding X and Y, then there must be a physical cause for this difference) but is one of epistemological procedure (If *we are going to maintain* different things regarding X and Y, then we must have *a good and sufficient reason* for doing so). And this principle is not constitutive but regulative, in Kant's terminology. It is not at all a *substantive* principle of the physical operation of nature but a fundamentally *economic* principle of rational procedure. For the advice, Go to the trouble of devising different claims only where this is necessitated by different circumstances, is a straightforward matter of rational economy—or simplicity of cognitive procedure.

It is useful to reexamine the Anaximander argument in this economic-epistemological light. What is now seen to be at issue with the sort of symmetry argument operative in Anaximander's explanation is not the idea that an object cannot break physical symmetry. Rather, the line of thought goes something like this: In the event that we just have no reason to claim the earth's position in the cosmos does break symmetry (i.e., since symmetry exists in all relevantly determinable respects), it would clearly be foolish of us to ask for a *further* explanation of symmetrical comportment. The interests of explantory economy are best achieved by letting the fact of apparent symmetry itself provide the needed account. What is at issue is thus not a physical principle of symmetry but a procedural principle of cognitive economy.

To be sure, the role played by symmetry considerations in such an invocation of the economy of cognition is merely presumptive. What is at issue is not a commitment to the very questionable claim that nature must or does not break symmetry, but rather a commitment to the economic principle that when things go in the normal way one does not ask for further explanations—that where no anomaly calls for an explanation, we leave matters alone and expend our cognitive efforts in other, more promising directions.

CONCLUSION

Methodologists of science have had a tendency to treat epistemological issues on the basis of an ideological attachment to such abstractly theoretical factors as generality, simplicity, testability, explanatory power, robustness, coherence, and novelty. The deliberations of this chapter indicate that a far more natural approach is available along pragmatic/economic lines. Once we bring to bear the usual machinery of rational decision making with its concern for the sensible balance of costs and benefits, then we need no longer argue directly, on independent substantive grounds, for the explanatory significance of these various parameters of epistemic virtue. We can readily provide a straightforward validation of the role of such factors in our methodology of inquiry on the standard principles of the rational economy of cognitive procedure. For the decisive factor is their overall cost effectiveness in facilitating the aims of the scientific enterprise in terms of its characteristic objectives: description, explanation, and control over nature. The legitimacy of standard cognitive values and processes can be adjudged within this overarching framework on the basis of the cost-effectiveness considerations at issue with our commitment to rational procedure in general.[56]

SEVEN

Cost Escalation in Empirical Inquiry

SYNOPSIS

(1) The development of inquiry in natural science is best understood on the analogy of exploration—not geographical exploration, to be sure, but exploration in the parametric space of such physical factors as temperature, pressure, and field strength. (2) This interactive exploration becomes increasingly difficult (and expensive) as we move ever further away from the home base of the accustomed environment of our evolutionary heritage. (3) As its data base grows, scientific theorizing involves the construction of the least complex theoretical structure to accommodate the available data. (4) Theory thus develops in a dialectical interchange with experimentation through an ongoing succession of equilibria and disequilibria. (5) The cost escalation governing the development of scientific technology sets fundamentally economic limits to the advance of natural science. (6) In sum, the economic dimension of empirical inquiry is an inherent structural feature of the enterprise, which theorists of

knowledge can overlook only at the cost of injury to their own endeavors.

THE EXPLORATION MODEL OF SCIENTIFIC INQUIRY

In developing natural science, we humans began by exploring the world in our own locality—not just our spatial neighborhood but our parametric neighborhood in the space of physical variables such as temperature, pressure, and electric charge. Near the "home base" of the state of things in our accustomed natural environment, we can operate with relative ease and freedom—thanks to the evolutionary heritage of our sensory and cognitive apparatus—scanning nature with the unassisted senses for data regarding its modes of operation. But in due course we accomplish everything that can be managed by these straightforward means. To do more, we have to extend our probes into nature more deeply, deploying increasing technical sophistication to achieve more and more demanding levels of interactive capability. We have to move ever further away from our evolutionary home base in nature toward increasingly remote frontiers. From the egocentric standpoint of our local region of parameter space, we have, over time, journeyed ever more distantly outward to explore nature's various parametric dimensions, in the manner of a prospector, searching for cognitively significant phenomena.

The appropriate picture is not, of course, one of geographical exploration but of the physical exploration—and subsequent theoretical systematization—of phenomena distributed over the parametric space of the physical quantities spreading out all about us. This approach in terms of exploration provides a conception of scientific research as a prospecting search for the new phenomena needed for significant new scientific findings. As the range of telescopes, the energy of particle accelerators, the effectiveness of low-temperature instrumentation, the potency of pressurization equipment, the power of vacuum-creating contrivances, and the accuracy of measurement apparatus increases—that is, as our capacity to move about in the parametric space of the physical world is enhanced—new phenomena come into view. The key to the great progress of contemporary physics lies in the enormous strides which an ever more sophisticated scientific technology enables us to make by enlarging the empirical basis of our knowledge of natural processes.[57]

No doubt, nature is in itself uniform as regards the distribution of its diverse processes across the reaches of parameter space. It does not favor us by clustering them in our accustomed parametric vicinity: significant phenomena do not dry up outside our parochial neighborhood. But cognitively significant phenomena in fact become increasingly sparse, because scientific ingenuity is able to do so much so well early on. Our power of theoretical triangulation is so great that we can make disproportionately effective use of the phenomena located in our local parametric neighborhood. But scientific innovation becomes more and more difficult—and expensive—as we push our explorations even further away from our evolutionary home base toward increasingly remote frontiers. After the major findings accessible at a given data technology level have been achieved, further major findings become realizable only when one ascends to the next level of sophistication in data-relevant technology.

TECHNOLOGICAL ESCALATION

Natural science is fundamentally empirical, and its advance is critically dependent not on human ingenuity alone but on the monitoring observations to which we can gain access only through interactions with nature. The days are long past when useful scientific data can be had by unaided sensory observation of the ordinary course of nature. Artifice has become an indispensable route to the acquisition and processing of scientifically useful data. The sorts of data on which scientific discovery nowadays depends can be generated only by technological means.

The enormous power, sensitivity, and complexity deployed in present-day experimental science have not been sought for their own sake but rather because the research frontier has moved on into an area where this sophistication is the indispensable requisite of ongoing progress. In science, as in war, the battles of the present cannot be fought effectively with the armaments of the past.

Without an ever-developing technology, scientific progress would soon grind to a halt. The discoveries of today cannot be made with yesterday's equipment and techniques. To conduct new experiments, to secure new observations, and to detect new phenomena, an ever more powerful investigative technology is needed.

Scientific progress depends crucially and unavoidably on our technical capability to penetrate into the increasingly distant—and increasingly difficult—reaches of the spectrum of physical parameters, to explore and to explain the ever more remote phenomena encountered there. we are embarked on a literally limitless endeavor to improve the range of effective experimental intervention, because only by operating under new and heretofore inaccessible conditions of observational or experimental systematization—attaining extreme temperature, pressure, particle velocity, field strength, and so on—can we realize those circumstances that enable us to put our hypotheses and theories to the test. As one acute observer has rightly remarked: "Most critical experiments [in physics] planned today, if they had to be constrained within the technology of even ten years ago, would be seriously compromised."[58] Throughout the natural sciences, technological progress is a crucial requisite for cognitive progress.

This idea of the exploration of parametric space provides a basic model for understanding the mechanism of scientific innovation in mature natural science. New technology increases the range of access within the parametric space of physical processes. Such increased access brings new phenomena to light, and the detection, examination, and theoretical systematization of these phenomena is the basis for growth in our scientific understanding of nature.

Such deliberations point toward the idea of technological levels corresponding to the successive state-of-the-art stages in the technology of inquiry with regard to data generation and processing, each giving rise to successively later generations of investigative instrumentation and machinery. These levels are generally separated from one another by substantial (order-of-magnitude) improvements in performance with regard to such information-providing parameters as measurement exactness, data processing volume, detection sensitivity, high voltages, and high or low temperatures.

The salient characteristic of this situation is that, once the major findings accessible at a given data-technology level have been attained, further major progress in the problem area requires ascent to a higher level on the technological scale. Every data technology level is subject to discovery saturation, but the exhaustion of prospects at a given level does not, of course, bring progress to a stop. Once the potential of a given data technology level has been exploited, not all our piety or wit can lure the ongoing frontier back to yield further significant returns at this stage. To be sure, after the major findings accessible at a given data-technology level have been realized, further major findings become realizable by ascending to the next level of sophistication in data-relevant technology.

We arrive therefore at the phenomenon of *technological escalation*. The need for new data forces us to look further and further from man's familiar home base in the parametric space of nature. Thus while scientific progress is in principle always possible—there being no absolute or intrinsic limits to significant scientific discovery—the realization of this ongoing prospect demands a continual enhancement in the technological state of the art of data extraction or exploitation.

This technological escalation has massive economic ramifications. The economics of scientific inquiry presents a picture of ongoing cost escalation that is strongly reminiscent of an arms race. The technical escalation inherent in scientific research parallels that familiar arms race situation of inescapable technological obsolescence, as the opposition escalates to the next phase of sophistication. In both cases, the economic structure is the same, as new technology leads to exponential cost increases. (Consider the series of the B bombers, from the old B-17 of World War II, through the B-47 and B-52 of the Cold War era, to the supersonic B-1 of today.) As science endeavors to extend its mastery over nature, it embarks on a technology-intensive arms race against nature. The escalation of technological capabilities—and, correlatively, of costs—is the manifestation of this phenomenon.

In any matured branch of natural science, continually greater capabilities in terms of technological capacity are required to realize further substantial results. Thus the purchase price of significant new findings constantly increases. Once all the significant findings accessible at a given state-of-the-art level of investigative technology have been realized, one must continually move on to a new, more complex (and thus more expensive) level: one requires more accurate measurements, more extreme temperatures, higher voltages, more intricate combinations, and so on. Over time, the ongoing requirement for manpower and material to sustain smooth progress has been increasing at an exponentially increasing rate. The phenomenon of cost escalation is explained through a combination of the finitude of the body of substantial results realizable at a given level of investigative technology, together with a continual and steep increase in the resource costs of pushing from one level to the next.

The perspective afforded by such an escalation process indicates that progress in natural science was at first relatively undemanding, because we have explored nature in our own parametric neighborhood.[59] Economic demands were minimal at this stage, because the requisite technology was relatively crude. But over time an ongoing escalation in the resource costs of significant scientific discovery arose from the increasing technical difficulties of realizing this objective, difficulties that are a fundamental—and an ineliminable—part of an enterprise of empirical research, for we must here contrive ever more "far out" interactions with nature, operating in a continually more difficult, accordingly, sector of parametric space.

THEORIZING AS INDUCTIVE PROJECTION

In pursuing the venture of scientific inquiry, we scan nature for interesting phenomena and for the explanatorily useful regularities suggested by them. As a fundamentally inductive process, scientific theorizing involves exactly this devising of the least complex theory structure capable of accommodating the available data. At each stage we try to embed the phenomena and their regularities within the simplest (cognitively most efficient) explanatory fabric to answer our questions about the world and to guide our interactions in it.

One very important point must, however, be stressed in this connection. The idea of a coordinative systematization of question-resolving conjecture with the data of experience may sound like a very conservative process. This impression would be quite incorrect. The drive to systematization embodies an imperative to broaden the range of our experience—to extend and to expand the data base from which our theoretical triangulations proceed—which is no less crucial than assuming their elegance. Simplicity/harmony and comprehensiveness/inclusiveness are two components of one whole. That is why the ever-widening exploration of nature's parameter space is an indispensable part of the process.

With the enhancement of scientific technology, the size and complexity of this body of data inevitably grows, expanding on quantity and diversifying in kind. Technological progress constantly enlarges the window through which we look out upon parametric space. In developing natural science, we use this window to scrutinize parametric space, continually augmenting our data base and then generalizing upon what we see. And what we have here is not a homogeneous lunar landscape, where once we have seen one sector we have seen it all, and where theory projections from lesser data generally remain in place when further data comes our way. Historical experience shows that there is every reason to expect that our ideas about nature are subject to constant radical changes as we explore parametric space more extensively. The technologically mediated entry into new regions of parameter space constantly destabilizes the attained equilibrium between data and theory.

Our exploration of physical parameter space is inevitably incomplete. We can never exhaust the whole of these parametic ranges of temperatures, pressures, particle velocities, and the like because of physical resistance as one moves toward the extremes. And so we inevitably face the (very real) prospect that the regularity structure of the as yet inaccessible cases generally does not conform to the patterns of regularity prevailing in the currently accessible cases. In general, new data just do not accommodate themselves to old theories. (Newtonian calculations, for example, worked marvelously for predicting solar system phenomenology—eclipses, planetary conjunctions, and the rest—but this did not mean that classical physics was free of any need for fundamental revision.)

Scientific theory formation is, in general, a matter of spotting a local regularity of the phenomena within a limited region of parametric space and then projecting globally across the board. The theoretical claims of science are not spatiotemporally localized, and they are not parametrically localized either. The theories of theoretical science stipulate how things are always and everywhere. And so it does not require a sophisticated knowledge of statistics to recognize the inductive projection of the sort we make in science is invariably a risky enterprise. And it does not require a sophisticated knowledge of the history of science to recognize that our worst fears are usually realized—that it is seldom if ever the case that our theories survive intact in the wake of extensions in our access to sectors of parametric space. Both as regards the observable *regularities* of nature and the discernible *constituents* of nature, very different results emerge at various levels of the observational state of the art. At every stage of investigative sophistication we seem to confront a different order or aspect of things. What we find in investigating nature must always in some degree reflect the character of our technology of observation. What we can detect, or find, in nature is always something that depends on the mechanisms by which we search. The history of science is thus a succession of episodes of leaping to the wrong conclusions, while yet doing so in a manner that is conscientious and responsible in its use of the available data.

Even extraordinary accuracy with respect to the entire range of currently manageable cases does not betoken actual correctness, it merely reflects adequacy over that limited range. And no matter how greatly we broaden that limited range of currently accessible cases, we still achieve no assurance (or even probability) that a theory corpus that smoothly accommodates the whole range of currently achievable outcomes will hold across the board. The prospect of future change—that is, improvement—can never be confidently foreclosed.

THE DIALECTIC OF EXPERIMENTATION AND THEORIZING

Given that we can only learn about nature by interacting with it, Newton's third law of countervailing action and reaction becomes a fundamental principle of epistemology. Everything depends on just how *and how hard* we can push against nature in situations of observational and detectional interaction. As Bacon saw, nature will never tell us more than we can forcibly extract from her with the means of interaction at our disposal. And what we can manage to extract by successively deeper probes is bound to wear a steadily changing aspect, because we operate in new circumstances where old conditions cannot be expected to prevail. Phenomenological novelty is seemingly inexhaustible: we can never feel confident that we have got to the bottom of it. Nature always has hidden reserves of phenomena at her disposal. Successive stages in the technological state of the art of scientific inquiry lead us to ever-different views about the nature of things and the character of their laws.

Progress in natural science is a matter of dialogue or debate between theoreticians and experimentalists. The experimentalists probe nature to see its reactions, to seek out phenomena. And the theoreticians take the resultant data and weave a theoretical fabric about them. Seeking to devise a framework of rational understanding, they construct their explanatory models to accommodate the findings that the experimentalists put at their disposal.

But once the theoreticians have had their say, the ball returns to the experimentalists' court. Employing new, more powerful means for probing nature, they bring new phenomena to view, new data for accommodation. Precisely because these data are new and inherently unpredictable, they often fail to fit the old theories. Theory extrapolations from the old data could not encompass them; the old theories do not accommodate them. And so a disequilibrium arises between existing theory and new data.

At this stage, the ball reenters the theoreticians' court. New theories must be devised to accommodate the new, nonconforming data. And so the theoreticians set about weaving a new theoretical structure to accommodate the new data. They endeavor to create, once more, an equilibrium between theory and data. And then the ball returns to the experimentalists' court, and the whole process starts over again.

But the ongoing escalation in the technological requisites for scientific progress means that each iteration gets more complex and more expensive. With the progress of science, nature becomes less and less yielding to the efforts of our inquiry. We are faced with the need to push nature harder and harder to achieve cognitively profitable interactions. The dialectic of theory and experiment carries natural science ever deeper into the range of greater costs. Science, in sum, involves something of a technological arms race against nature.

THE ECONOMIC LIMITS OF SCIENCE

While we can confidently anticipate that our scientific technology will undergo ongoing improvement, we cannot expect it ever to attain perfection. There is no reason to think that we ever will, or indeed can, reach the end of the road here. Every successive level of technical capability has its inherent limitations, whose overcoming calls for achieving yet another, more sophisticated, level of the technological state of the art.

The intensity of pressure and temperature can in principle always be increased, our low-temperature experiments brought closer to absolute zero, atomic particles accelerated ever closer to the speed of light, vacua made more and more perfect, and so on. There is always more to be done, because interest resistance limits perfection. And experience teaches that any such enhancement of physical mastery brings new phenomena to view, providing an enhanced capability to test yet further hypotheses and to discriminate between new alternative theories to enhance our knowledge of nature.

There will always be interactions with nature on a scale whose realization requires the deployment of greater resources than we have heretofore expended. But man's material resources are limited. These limits inexorably circumscribe our experiential access to the real world. And when there are interactions to which we have no access, we must presume phenomena that we cannot discern. It would be very odd indeed if nature were to confine the distribution of scientifically significant phenomena to those ranges that happen to lie conveniently within our reach—a condition counterindicated by the whole course of our prior experience.

Given the limitations of our access to nature's phenomena, we must come to terms with the fact that we cannot realistically expect that our science will ever, at any given stage of its development, be in a position to afford us more than a partial and incomplete account of nature, for the achievement of cognitive control over nature requires not only intellectual instrumentalities (concepts, ideas, theories, knowledge) but also, and no less importantly, the deployment of physical resources (technology and power). And since the physical resources at our disposal are restricted and finite, it follows that our capacity to effect control is bound to remain imperfect and incomplete, with much in the realm of the doable always remaining undone. We shall never be able to travel down this route as far as we might like to go.

Progress in modern natural science faces the challenging task of climbing ever upward from one level of technological sophistication to the next. In natural science, creative genius cannot of itself outrun the course of technological development. The Danish historian of science A. G. Drachmann closed his excellent book *The Mechanical Technology of Greek and Roman Antiquity* with the following observation: "I should prefer not to seek the cause of the failure of an invention in the social conditions till I was quite sure that it was not to be found in the technical possibilities of the time."[60] The history of science, as well as that of technology, is crucially conditioned by the limited nature of "the technical possibilities of the time."

Natural science being an inescapably empirical enterprise, remorseless limitations are imposed upon the prospects of effective theorizing at any given stage in its development by its dependency on the data-generative technology of the day. To say this is not to sell human ingenuity short; it is simply a matter of facing a very fundamental fact of scientific life. Progress in natural science is insuperably limited at any given time by the implicit barriers set by the available technology of data acquisition and processing. *Technological dependency sets technological limits*—first to data acquisition and then to theory projection. The achieved level of sophistication in the technological state of the art of information acquisition and processing sets definite limits to the prospects of scientific progress by restricting the range of findings that are going to be realistically accessible.

One of the clearest lessons of the history of science is that as we acquire more powerful means of data acquisition and processing, and thus as our information base changes, so the character of our theories, and with it our view of the world—our "picture" of nature—also changes. The existence of a potentially unending sequence of levels of technological sophistication entails an unending (potential) sequence of levels of theoretical sophistication, with a very different story, a different picture of nature, emerging at every level.

Progress without new data is, of course, possible in various fields of scholarship and inquiry. The example of pure mathematics, for instance, shows that discoveries can be made in an area of inquiry that operates without empirical data. But this hardly represents a feasible prospect for natural science. It is exactly the explicit dependency on additional data—the empirical aspect of the discipline—that sets natural science apart not only from the formal sciences (logic and mathematics) but also from the hermeneutic ones (like the humanities), which address themselves ceaselessly to the imaginative reinterpretation and re-reinterpretation of old data from novel conceptual perspectives.

The ancient Greeks were certainly as intelligent as we

are—perhaps even more so. But given the information technology of the day, it is not just improbable but actually inconceivable that the Greek astronomers could have come up with an explanation for the red shift or the Greek physicians with an account of the bacteriological transmission of some communicable disease. The relevant types of data needed to put such phenomena within cognitive reach simply lay beyond their range. Given the instrumentalities of the times, there just was no way for the Greeks (no matter how well endowed with brain power) to gain physical or conceptual access to the relevant phenomena. Progress in theorizing in these directions was barred, not permanently but *for them*, by a technological barrier of access of data, a barrier as absolute as the then-extant technological barriers in the way of developing the internal combustion engine or the wireless telegraph.

There remains an insidiously tempting argument: the contention that a slowing in access to new phenomena under the retarding impetus of (increasingly significant) resource limitations doesn't matter all that much, because these further capabilities would in any case afford no more than icing on the cake, trivial refinements and minor corrections located more decimal places further on. The view that underlies such a position is that further changes are smaller changes, that a juncture has been reached where the additional advances of science are merely minor details and readjustments in a basically completed picture of how nature functions.

This unrealistically optimistic view deserves outright rejection. There is simply no reason to think that nature conveniently assures that phenomena of difficult access are thereby insignificant in cognitive importance, so that we come, early and easily, near to exhausting the range of cognitively significant interactions. We cannot realistically expect that our science, at any given stage of its development, will ever be in a position to afford us more than a very partial and incomplete degree of cognitive and physical control over nature. The reach of our interactions with nature—and thus, presumably, the reach of our cognitive control of nature—remains imperfect and incomplete. And in natural science, imperfect physical control is bound to mean imperfect cognitive control.

Limitations of physical capacity and capability spell cognitive limitations for empirical science. Where there are inaccessible phenomena, there must be cognitive inadequacy as well. To this extent, at any rate, the empiricists were surely right. Only the most fanatical rationalist could uphold the capacity of sheer intellect to compensate for lack of data. The existence of unobserved phenomena means that our theoretical systematizations may well be (and presumably are) incomplete. Insofar as certain phenomena are not just undetected but in the very nature of the case inaccessible (even if only for the merely economic reasons suggested above), our theoretical knowledge of nature must be presumed imperfect. Fundamental features inherent in the structure of man's interactive inquiry into the ways of the world thus conspire to ensure the incompleteness of our scientific knowledge of nature.

In the end, then, natural science confronts not barriers (boundaries or absolute limits) but obstacles (difficulties and impediments). The technological/economic requirements for scientific progress mean that we will never be able to advance the project to our total satisfaction—will never be able to do as much as we would, ideally speaking, like to do. Given the inescapable realities of resource limitations, the prospect of a perfected or complrte science of nature is a practical impossibility. In a world of limited resources, science could certainly be ended—finished in the sense of being advanced as far as it is possible for creatures of our kind to push a venture of this sort—without thereby being perfected or completed, that is, without discharging to the full the characterizing mandate of the enterprise in terms of description, explanation, prediction, and control.

Further progress becomes increasingly difficult, not because we have come close to completing science—to exhausting nature's stock of new phenomena—but because the circumstance that there are indeed rich veins deeper down in the mine avails us only if we can actually dig there. It is not that we have managed to get to the the bottom of things but that we have carried the process as far as we can afford to—as far as the means at our disposal allow us to go. As best we can tell, the limits of science are economic, we reach them not because we have exhausted the novelties of nature but because we have come to the end of our economic tether. It gets increasingly difficult—and ultimately, in a world of limited resources, just too expensive—to push forward the frontiers of technology. So also does it get increasingly expensive to advance the frontiers of theorizing.

The limits of science are very real, but they are not inherently intellectual matters of human incapacity or deficient brain power. They are fundamentally economic limits imposed by the technological character of our access to the phenomena of nature. The overoptimistic idea that we can push science ever onward to the solution of all questions that arise shatters in the awkward reality that the price of problem solving inexorably increases to a point beyond the limits of affordability.[61]

CONCLUSION

It has been a serious shortcoming of philosophical epistemology that it has to date focussed on product to the neglect of process. For inquiry—in science and elsewhere—is a human activity which, like any other, requires the expenditures of effort and energy in a way that endows the enterprise with an unavoidable economic dimension. Economic factors shape and condition our cognitive proceedings in so fundamental a way that they demand explicit attention. The lesson of the book is thus straightforward: only by heeding the concrete processes that engender our knowledge in a way that takes account of their economic dimension can we adequately explain the nature of its operations and properly understand the character of its products. As we have seen repeatedly, a cogent account for our standard cognitive practices is provided by seeing them as emerging from an economic pressure to cost effectiveness in the management of our epistemic affairs.

Currently, fashionable relativism to the contrary notwithstanding, our standard ways of conducting inquiry are not simply a matter of conforming to the prevailing fashions and established customs of the time and place. Rather, they serve as a justifying basis of rational legitimation once we regard them in problem-solving terms, for they can in substantial measure, be justified in essentially pragmatic terms as representing cost-effective means to characterizing each epistemic enterprise—the securing of viable answers to our questions about the world. The best explanation for many or most of our cognitive practices is simply—and prosaically—that it makes good rational sense for them to be as they are.

NOTES

INDEX

NOTES

1. On Peirce's project on economy of research, see the author's *Peirce's Philosophy of Science* (Notre Dame and London, 1976), as well as C. F. Delaney, "Peirce on 'Simplicity' and the Conditions of the Possibility of Science," in L. J. Thro, ed., *History of Philosophy in the Making* (St. Louis, 1974), pp. 177–94.

2. One valuable contribution in this area is Fritz Machlup, *The Production and Distribution of Knowledge in the United States* (Princeton, 1962).

3. Fridtjof Nansen as quoted in Roland Huntford, *The Last Place on Earth* (New York, 1985), p. 200.

4. William James, "The Sentiment of Rationality," in *The Will to Believe and Other Essays in Popular Philosophy* (New York and London, 1897), pp. 78–79.

5. But is it indeed irrational to give a gift more costly than the social situation requires? By no means! It all depends on one's aims and ends, which may, on such an occasion, lie in a desire to cause the recipient surprise and pleasure, rather than merely doing the customary thing.

6. On this theme, see the important investigations of George K. Zipf, *Human Behavior and the Principle of Least Effort* (Boston, 1949). Zipf's investigations furnish a wide variety of interesting examples of how various of our cognitive proceedings exemplify a tendency to minimize the expenditure of energy.

7. As William James said: "[Someone] who says 'Better to

go without belief forever than believe a lie!' merely shows his own preponderant private horror of becoming a dupe.... But I can believe that worse things than being duped may happen to a man in this world" (*The Will to Believe,* pp. 18–19).

8. H. H. Price, *Belief* (London, 1969), p. 128.
9. David Hume, *An Enquiry Concerning Human Understanding,* sec. XII, pt. ii. Compare John Locke: "He that will not eat till he has a demonstration that it will nourish him; he that will not stir till he infallibly knows the business he goes about will succeed, will have little else to do but to sit still and perish" (*Essay Concerning Human Understanding,* bk. IV, chap. XIV, sec. 1).
10. Sextus Empiricus, *Outlines of Pyrrhonism,* bk. I, chap. 20, sec. 193; and compare secs. 121–24.
11. Ludwig Wittgenstein, *On Certainty* (Oxford, 1969), sec. 287 (italics supplied).
12. On priority conflicts, see R. K. Merton, "Priorities in Scientific Discovery," *American Sociological Review* 22 (1957): 635–59.
13. The founding of the Royal Society in London, chartered in 1666, was a small but significant step toward openness. For an interesting analysis of the historical situation, see Jerome R. Ravetz, *Scientific Knowledge and Its Social Problems* (Oxford, 1971). Ravetz's presentation makes it clear that a considerable transformation was involved because "a significant proportion of the great 'scientists' of that age [sixteenth to seventeenth century] were even more concerned for the protection of their intellectual property, than for an immediate realization of its value through the prestige resulting from publication" (p. 249).

14. The situation is one of the sort called a prisoner's dilemma by game theorists. For a good account, see Morton D. David, *Game Theory* (New York; 1970), pp. 92–103. See also A. Rapoport and A. M. Chammah, *Prisoner's Dilemma: A Study in Conflict and Cooperation* (Ann Arbor, 1965), and Anatol Rapoport, "Escape from Paradox," *American Scientist* 217 (1967): 50–56.

15. Compare H. M. Vollmer and D. L. Mills, eds., *Professionalization* (Englewood Cliffs, 1966). This credit, once earned, is generally safeguarded and maintained by institutional means: licensing procedures, training qualifications, professional societies, codes of professional practice, and the like.

16. On this matter, see Thomas Sowell, *Knowledge and Decisions* (New York, 1980), especially the discussion of "Informal Relationships," on pp. 23–30.

17. For a penetrating study of these and similar issues, see John Sabini and Maury Silver, *Moralities of Everyday Life* (Oxford, 1982), especially chap. 4.

18. Specifically, Solomon Asch found that in certain situations of interactive estimation, "whereas the judgments were virtually free of error under control conditions, one-third of the minority estimates were distorted toward the majority." See his "Studies of Independence and Conformity: I. A Minority of One against a Unanimous Majority," *Psychological Monographs: General and Applied*, no. 70 (1956).

19. Ibid, p. 69.

20. John Sabini and Maury Silver, *Moralities of Everyday Life*, pp. 84–85.

21. Harry Kalven, Jr., and Hans Zeisel, *The American Jury* (Chicago, 1966).

22. Usefully relevant discussions can be found in David Lewis, *Convention: A Philosophical Study* (Cambridge, 1969). But contrast Angus Ross, "Why Do We Believe What We are Told?" *Ratio* 28 (1986): 69–88.

23. H. P. Grice, "Meaning," *The Philosophical Review* 66 (1957): 377–88. Compare also Jonathan Bennett, *Linguistic Behavior* (London, 1963), chaps. 1 and 7.

24. Norman Storer's study, *The Social System of Science* (New York, 1966), projects the model of an exchange operated by the scientific community that trades the rewards of recognition for the creative effort of the scientist.

25. Vagueness constitutes a context in which we trade off informativeness (precision) with probable correctness (security), with science moving toward the former, and everyday knowledge toward the latter. The relevant issues are considered in tantalizing brevity in Charles S. Peirce's short discussion of the logic of vagueness, which he laments as too much neglected, a situation that has since been corrected only partially.

26. Jerome R. Ravetz, *Scientific Knowledge and its Social Problems* (Oxford, 1971), p. 244.

27. Strange though it may seem, importance is a neglected topic in philosophy. Little, if any, treatment of the issue can be found in the literature. No philosophical dictionary or encyclopedia I know of has an article on importance. The excellent *Diccionario de Filosofía* of José Ferrater Mora (Madrid, 1979–80) does recognize the existence of the idea by carrying the entry "IMPORTANCIA:

véase RELEVANCIA." But that of course is something else again.

28. Nicholas Maxwell, *From Knowledge to Wisdom: A Revolution in the Aims and Methods of Science* (Oxford, 1984).

29. Ibid, p. 111.

30. Shakespeare saw the matter aright; what he says of worth in general certainly holds for cognitive worth or importance: "But value dwells not in particular will / It holds its estimate and dignity / As well wherein 'tis precious of itself / As in the prizer" (*Troilus and Cressida*, act 2, sc. 2, lines 53–56).

31. Larry Laudan is one of the few writers on the philosophy of science who recognize that, since the answering of important questions and the resolution of important problems is the object of the scientific enterprise, an adequate theory of science must address the issue of importance. Unfortunately, however, he speaks of "interesting questions [or] in other words . . . important problems" (Larry Laudan, *Progress and Its Problems* [Berkeley, 1977], p. 13). And this is highly problematic. In science as elsewhere, issues can be interesting without necessarily thereby being very important—for example, explaining the extinction of dinosaurs.

32. William Whewell, *Novum Organon Renovatum* (London, 1858), p. 114.

33. The force of Dickinson Miller's principle must be acknowledged: "There is no intermediate degree between following from premises and not following from them. There is no such thing as half-following or quarter-following" (Dickinson S. Miller, "Professor Don-

ald Williams vs. Hume," *The Journal of Philosophy* 44 (1947): 673–84 (see p. 684).

34. This perspective supports F. H. Bradley in his critique of J. S. Mill's view of induction, on the basis of the consideration that inference as such is impotent to accomplish the move from particulars to universals because it is only legitimate to argue from some to all if it is premissed that the particulars at issue share some universal character.

35. Carl G. Hempel, *Philosophy of Natural Science* (Englewood Cliffs, 1966), p. 15.

36. A. A. Cournot, *Essai sur les fondements de nos connnaisances*, vol. 1 (Paris, 1851), p. 82.

37. Compare John G. Kemeny, "The Use of Simplicity in Induction," *The Philosophical Review* 62 (1953): 391–408.

38. Henri Poincaré, *Science and Hypothesis* (New York, 1914), pp. 145–46.

39. Galileo Galilei, *Dialogues Concerning Two New Sciences*, trans. H. Crew and A. de Salvo (Evanston, 1914), p. 154.

40. Kant was the first philosopher clearly to perceive and emphasize this crucial point:

    But such a principle [of systematicity] does not prescribe any law for objects; . . . it is merely a subjective law for the orderly management of the possessions of our understanding, that by the comparison of its concepts it may reduce them to the smallest possible number; it does not justify us in demanding from the objects such uniformity as will minister to the convenience and extension of our understanding; and we may not, therefore, ascribe to the [methodological or regulative] maxim [Systematize knowledge!] any objective [or descriptively *constitutive*] validity. (CPuR., A306 = B362.)

Compare also C. S. Peirce's contention that the systematicity of nature is a regulative matter of scientific attitude rather than a constitutive matter of scientific fact. Charles Sanders Peirce, *Collected Papers*, vol. 7, sec. 7.134.

41. Some issues revolving around this principle are discussed in Daniel Goldstick, "Methodological Conservatism," *American Philosophical Quarterly* 8 (1971): 186–91.

42. To be sure, these two factors can come into conflict, in which case one must balance things out. (See note 45.)

43. Hans Reichenbach, *Experience and Prediction* (Chicago and London, 1938), p. 376. Compare:

    Imagine that a physicist . . . wants to draw a curve which passes through [points on a graph that represent] the date observed. It is well known that the physicist chooses the simplest curve; this is not to be regarded as a matter of convenience [for different] curves correspond as to the measurements observed, but they differ as to future measurements; hence they signify different predictions based on the same observational material. The choice of the simplest curve, consequently, depends on an inductive assumption: we believe that the simplest curve gives the best predictions. . . . If in such cases the question of simplicity plays a certain role for our decision, it is because we make the assumptions that the simplest theory furnishes the best predictions. (Ibid., pp. 375–76.)

44. See B. W. Petky, *The Fundamental Physical Constants and the Frontiers of Measurement* (Bristol and Boston, 1985).

45. These considerations explain how we are to proceed in situations where the parameters of cognitive systemati-

zation stand in apparent conflict with one another, when conformity seems at odds with cohesiveness, or the like. Reconciliation is to be effected here in terms of the demands of *overall* economy.

46. Further considerations relevant to these issues are canvassed in the author's *Methodological Pragmatism* (Oxford, 1977) and *Cognitive Systematization* (Oxford, 1979).

47. C. G. Hempel, "A Note on the Paradoxes of Confirmation," *Mind* 55 (1946): 79–82. See also Rudolf Carnap, *Logical Foundations of Probability*, 2d ed. (Chicago, 1962), pp. 223–24.

48. For a survey of positions and objections to them, see Henry E. Kyburg, Jr., "Recent Work on Inductive Logic," *American Philosophical Quarterly* 1 (1964): 249–78. Kyburg summarizes his discussion with the observation: "The problem of finding some way of distinguishing between sensible predicates like 'blue' and 'green' and the outlandish ones suggested by Goodman, Barker, and others, is surely one of the most important problems to come out of recent discussions of inductive logic" (p. 266).

49. This idea of invoking ostension as a device for tackling Goodman's paradox is originally due to Wesley C. Salmon, "On Vindicating Induction," in H. E. Kyburg, Jr., and E. Nagel, eds., *Induction: Some Current Issues* (Middletown, 1963), pp. 27–41.

50. As Israel Scheffler—perhaps the ablest of Goodman's expositors and defenders—has remarked: "The most natural objection to Goodman's new approach is that it provides no explanation of entrenchment itself." See

Scheffler, "Inductive Inference: A New Approach," *Science* 27, (1958): 177–81).

51. Karl R. Popper, *The Logic of Scientific Discovery* (New York, 1959), p. 273.

52. Looking on theories as intellectual instrumentalities (for explanation, prediction, etc.), one can apply to them the usual economic consideration of many-sided versatility versus case-specific power that applies to tools in general. On the economic aspect of this latter issue, compare the suggestive discussion, "The Law of Diminishing Returns in Tools," in George Kingsley Zipf, *Human Behavior and the Principle of Least Effort* (Boston, 1949), pp. 66 ff., and also pp. 182 ff.

53. J. M. Keynes wrote in his *Treatise in Probability* (London, 1921) that "It is difficult to see, however, to what point the strengthening of an argument's weight by increasing the evidence ought to be pushed," since "there clearly comes a point when it is no longer worthwhile to spend trouble . . . in the acquisition of further information, and there is no evident principle by which to determine how far we ought to carry our maxim of strengthening the weight of [evidence]. A little reflection will probably convince the reader that this is a very confusing problem" (pp. 83–84). A more recent author concluded that "The question of when to stop gathering information is . . . one that has been considered by relatively few statisticians and almost no philosophers" (Henry E. Kyburg, *Probability and Inductive Logic* [London, 1920], p. 169). For a treatment of some relevant issue, see Michael E. Brady, "J. M. Keynes' 'Theory of Evidential Weight'," *Synthese* 71 (1987): 37–59.

54. Hippolytus, *Refutations*, I, 6; quoted in G. S. Kirk and

J. E. Raven, *The Presocratic Philosophers* (Cambridge, 1957), pp. 106–07. Compare Plato, *Phaedo*, 118E.

55. On the role of symmetry principles in natural science, see Hermann Weyl, *Symmetry* (Princeton, 1952); Eugene Wigner, *Symmetries and Reflections* (Bloomington, 1967); Anthony Zee, *Fearful Symmetry* (New York, 1986).

56. This chapter draws on the author's *Peirce's Philosophy of Science* (Notre Dame and London, 1978).

57. A homely fishing analogy of Eddington's is useful here. He saw the experimentalists as akin to a fisherman who trawls nature with the net of his equipment for detection and observation. Now suppose (says Eddington) that a fisherman trawls the seas using a fishnet of two-inch mesh. Then fish of a smaller size will simply go uncaught, and those who analyze the catch will have an incomplete and distorted view of aquatic life. The situation in science in the same. Only by improving our observational means of trawling nature can such imperfections be mitigated. (See A. S. Eddington, *The Nature of the Physical World* [New York, 1928]).

58. D. A. Bromley et al. *Physics in Perspective*, Student Edition (Washington, D.C., 1973); pp. 16, 23. See also Gerald Holton, "Models for Understanding the Growth and Excellence of Scientific Research," in Stephen R. Graubard and Gerald Holton, eds., *Excellence and Leadership in a Democracy* (New York, 1962), p. 115.

59. Note, however, that an assumption of the finite dimensionality of the phase space of research-relevant physical parameters becomes crucial here. For if these were limitless in number, one could always move on to the inexpensive exploitation of virgin territory.

60. A. G. Drachman, *The Mechanial Technology of Greek and Roman Antiquity* (Copenhagen and Madison, 1963). p. 206.

61. The present chapter draws on the author's *Scientific Progress* (Oxford, 1978), and also on *The Limits of Science* (Berkeley, Los Angeles, London; 1984).

NAME INDEX

Anaximander of Miletus, 126–28
Aristotle, 67
Asch, Solomon, 44, 157n18

Bacon, Francis, 142
Barker, S. F., 162n48
Becquerel, Henri, 76
Bennett, Jonathan, 158n23
Bradley, F. H., 160n34
Brady, Michael E., 163n53
Bromley, D. A., 164n58

Carnap, Rudolf, 162n47
Chammah, A. M., 157n14
Cournot, A. A., 86, 160n36

David, Morton D., 157n14
Delaney, C. F., 155n1
Drachmann, A. G., 145, 165n60

Eddington, A. S., 164n57
Einstein, Albert, 76
Empiricus, Sextus, 25, 156n10

Feyerabend, Paul K., 17

Galilei, Galileo, 92, 160n39
Goldstick, Daniel, 161n41
Goodman, Nelson, 108, 113–15, 117, 162n48, n50
Graubard, Stephen R., 164n58

Grice, H. P., 54, 158n23
Hempel, Carl G., 108–09, 160n35, 162n47
Holton, Gerald, 164n58
Hume, David, 24–25, 156n9
Huntford, Roland, 155n3

James, William, 7, 24, 155n4, n7

Kalven, Jr., Harry, 158n21
Kant, Immanuel, 160n40
Kemeny, John S., 160n37
Keynes, J. M., 108, 125, 163n53
Kirk, G. S., 163n54
Kyburg, Jr., Henry E., 162n48, n49, 163n53

Laudan, Larry, 159n31
Lewis, David, 158n22

Machlup, Fritz, 155n2
Maxwell, Nicholas, 159n28
Merton, R. K., 156n12
Mill, J. S., 160n34
Miller, Dickenson, 159n33
Mills, D. L., 157n15
Mora, José Ferrater, 158n27

Nagel, Ernest, 162n49
Nansen, Fridtjof, 7, 155n3

Newton, Isaac, 34

Peirce, Charles Sanders, 5–6, 123, 155n1, 158n25, 161n40, 164n56
Petky, B. W., 161n44
Plato, 164n54
Poincaré, Henri, 89, 160n38
Popper, Karl, 118, 121, 163n51
Price, H. H., 21, 156n8

Rapoport, Anatol, 157n14
Raven, J. E., 164n54
Ravetz, Jerome R., 156n13, 158n26
Reichenbach, Hans, 100, 106, 161n43
Ross, Angus, 158n22

Sabini, John, 44, 157n17, n20
Salmon, Wesley C., 162n49
Scheffler, Israel, 162n50
Shakespeare, William, 159n30
Silver, Maury, 45, 157n17, n20
Sowell, Thomas, 157n16
Storer, N., 158n24

Thro, L. J., 155n1

Vollmer, H. M., 157n15

Weyl, Hermann, 164n55
Whewell, William, 85, 159n32
Wittgenstein, Ludwig, 156n11

Zee, Anthony, 164n55
Zeisel, Hans, 158n21
Zipf, George K., 155n6, 163n52

www.ingramcontent.com/pod-product-compliance
Lightning Source LLC
Chambersburg PA
CBHW031249290426
44109CB00012B/493